New York Common Council

Railroads and Railroad Grants in the City of New York

New York Common Council

Railroads and Railroad Grants in the City of New York

ISBN/EAN: 9783744729451

Printed in Europe, USA, Canada, Australia, Japan

Cover: Foto ©berggeist007 / pixelio.de

More available books at **www.hansebooks.com**

RAILROADS

AND

RAILROAD GRANTS

IN THE

CITY OF NEW YORK.

SECOND AVENUE RAILROAD.

This Agreement, made this fifteenth day of December, in the year one thousand eight hundred and fifty-two, between the Mayor, Aldermen and Commonalty of the city of New York, parties of the first part, and Denton Pearsall, Joseph C. Skaden, Abraham B. Rapelyea, William L. Hall, Richard T. Mulligan, Charles Miller, Daniel J. Sherwood, Abraham Allen and Henry Goff, parties of the second part, being the persons named in the resolutions hereinafter set forth, to whom permission is given to lay or build a railroad track, in accordance with said resolutions.

Whereas, The said parties of the first part, in Common Council convened, did, on the eleventh day of December, one thousand eight hundred and fifty-two, duly pass and adopt the following resolutions, which that day became operative and binding, in the words and figures following:

Resolved, That permission is hereby given to Denton Pearsall, Joseph C. Skaden, Abraham B. Rapelyea, Wm. L. Hall, R. T. Mulligan, Charles Miller, Daniel J. Sherwood, Abraham Allen and Henry Goff, to lay a grooved railroad track in the following streets and avenues of the city of New York, viz.: Commencing at a point in the 2d avenue, at or near to 42d street, thence running down the 2d avenue to 23d street, with a double track, through 23d street, with a single track, to the 1st avenue; down 1st avenue to Allen street, through Allen street to Grand street, through Grand street to the Bowery, down the Bowery to Chatham street, across Chatham street to Oliver street, through Oliver street to South street, through South

street to Roosevelt street, across Roosevelt street to Front
street, through Front street to Peck slip, the terminus.
Returning, with a single track, as follows: Through Peck
slip to Pearl street, through Pearl street to Chatham street,
through Chatham street to the Bowery, through the Bowery to Grand street, through Grand street to Chrystie
street, through Chrystie street to the 2d avenue, to 23d
street, where it intersects the double track, and so on to its
termination, opposite the Harlem river, with a double
track.

Provided, however, That all the said rails shall be laid
down in such manner, and in such parts of the said streets
and avenues, as shall be approved by the Street Commissioner, so as to cause no impediment to the common and
ordinary use of the streets and avenues for all other purposes ; and that the water courses of the streets shall be
left free and unobstructed, and that the said company shall
pave the streets in and about the rails in a permanent manner, and keep the same in repair to the entire satisfaction
of said Street Commissioner ;

And Provided, further, That no motive power, except
horses, be used below 42d street; and further, that they run a
car on said road for the convenience of public travel, each and
every day, both ways, as often as every 15 minutes, from 5
to 6 o'clock, A. M., every 4 minutes, from 6 A. M. to 8 P. M.,
every 15 minutes, from 8 P. M. to 12 o'clock, P. M., and
every 30 minutes, from 12 o'clock P. M. to 5 o'clock, A. M.,
and as much oftener as public convenience may require,
under such directions as the Common Council may from
time to time prescribe.

Also, that the rate of passage on said railroad shall not

exceed a greater sum than 5 cents to 42d street, and also that the Common Council shall have power to regulate the fare for the entire length of said road, when it shall be completed to Harlem river.

Also, that said road shall be commenced within six months, and completed to 42d street within one year, and from 42d street to Harlem river within three years from the passage of this resolution.

Resolved, That the said parties shall, in all respects, comply with the direction of the Street Commissioner and of the Common Council, in the building of said railroad, and in the running of the cars thereon, and in any other matters connected with the regulation of said railroad.

Resolved, That the said parties shall, before this permission takes effect, enter into a good and sufficient agreement with the Mayor, Aldermen, and Commonalty of the City of New York, to be drawn and approved by the Counsel to the Corporation, binding themselves to abide by and perform the stipulations and provisions herein contained ; and also all such other regulations or ordinances as may be passed by the Common Council, relating to the said road.*

Now, therefore, this indenture witnesseth, that the said parties of the first part do make and declare this grant, and all licenses, rights and privileges and powers conferred or provided for, or intended to be conferred or provided for, in said resolutions, conditioned and dependent upon

---

* Adopted by the Board of Assistants, Nov. 9, 1852.

Adopted by the Board of Aldermen, Nov. 15, 1852.

Received from His Honor the Mayor, Dec. 11, 1852, without his approval or objection thereto ; therefore, under the provisions of the amended charter, the same became adopted.

the strict observance, performance and fulfilment by the said parties of the second part, or such of them as may act, of the said resolutions and the stipulations, restrictions, regulations and provisions therein contained, and the due and faithful performance by the said parties of the following covenants and agreements: That the said parties of the second part, for themselves and their successors, do hereby covenant and agree with the said parties of the first part, and with each other, that they will well and truly observe, perform, fulfill and keep the said resolutions hereinbefore particularly set forth, and all and every the provisions, stipulations, restrictions and conditions therein contained and thereby imposed, according to the true intent and meaning thereof.

In witness whereof, to one of these presents, remaining with the said parties of the first part, the said parties of the second part have set their hands and seals ; and to the other part thereof, remaining with the said parties of the second part, the said parties of the first part have caused the common seal of the city of New York to be affixed, the day and year first above written.

DENTON PEARSALL,	[L. S.]
ABRA'M ALLEN,	[L. S.]
RICHARD T. MULLIGAN,	[L. S.]
ABRA'M B. RAPELYEA,	[L. S.]
WM. L. HALL,	[L. S.]
CHAS. MILLER,	[L. S.]
DANIEL J. SHERWOOD,	[L. S.]
HENRY GOFF,	[L. S.]
JOSEPH C. SKADEN,	[L. S.]

Sealed and delivered in presence of

CHAS. H. HAWKINS.

CITY AND COUNTY OF NEW YORK, *ss.* :—On this 17th day of December, A. D. 1852, before me came Charles H Hawkins, subscribing witness to the within grant, to me personally known, who, being by me duly sworn, did depose and say that he is a resident of the city of New York, that he is acquainted with Denton Pearsall, Joseph C. Skaden, Abraham B. Rapelyea, William L. Hall, Richard T. Mulligan, Charles Miller, Daniel J. Sherwood, Abraham Allen and Henry Goff, and knows them to be the same individuals described in and who executed the within grant ; that he saw them sign the same ; that they severally acknowledged in his presence that they executed the same, and that he subscribed his name as a witness thereto.   GEORGE L. TAYLOR,

Commissioner of Deeds.

Resolved, That the route of the 2d Avenue Railroad be, and the same is hereby, changed from Front street, between Roosevelt street and Peck Slip, to South street, between the same points.

Adopted by the Board of Assistants, July 12, 1853.

Adopted by the Board of Aldermen, July 18, 1853.

Approved by the Mayor, July 20, 1853.

Resolved, That the 2d Avenue Railroad Company be, and are hereby directed to cause Grand street, from the Bowery to Allen street, to be repaired according to the terms of their grant, within 10 days from the adoption of this resolution.

Adopted by the Board of Councilmen, Feb. 13, 1854.

Adopted by the Board of Aldermen, April 3, 1854.

Approved by the Mayor, April 6, 1854.

Resolved, That the 2d Avenue Railroad Company be, and are hereby directed to pay to the City, for all pavement now laid and about to be laid in Grand street, in and about their tracks, as called for in their grant, and also Allen, between Grand and Broome streets.

Adopted by the Board of Aldermen, July 13, 1854.

Adopted by the Board of Councilmen, Oct. 6, 1854.

Approved by the Mayor, Oct. 11, 1854.

Resolved, That the 2d Avenue Railroad Company forthwith remove that portion of their track which extends from the southerly side of 49th street to the northerly side of 53d street, or so much thereof as shall, from time to time, be directed by the Street Commissioner, and that, if the said Railroad Company shall neglect or refuse so to remove the same within two days after being required so to do by the Street Commissioner, that said Street Commissioner be authorized and directed to remove the same.

Adopted by the Board of Councilmen, July 10, 1856.

Adopted by the Board of Aldermen, Aug. 4, 1856.

Approved by the Mayor, Aug. 6, 1856.

Whereas, An act passed the recent Legislature, authorizing the 2d Avenue Railroad Company to lay rails in certain streets of the city, other than those contained in their grant or permission from the Common Council, therefore, be it

Resolved, That should the said 2d avenue Railroad Company undertake to lay rails in any of the streets of the city, by the authority thus conferred upon said Company by the said Legislature, the Counsel to the Corporation is hereby authorized and directed to restrain said Company by injunction, and further, if necessary, to test the validity of said act.

By the Board of Aldermen, April 28, 1857.

By the Board of Councilmen, May 25, 1857.

Approved by the Mayor, May 27, 1857.

The President of the Croton Aqueduct Board directed to notify the 2d Avenue Railroad Company to put in good repair forthwith all the pavements in and about their rails, and, in case of neglect or refusal, to have the same done at the expense of the 2d Avenue Railroad Company.

Adopted by the Board of Aldermen, September 10, 1857.

Adopted by the Board of Councilmen, September 14, 1857.

Approved by the Mayor, September 16, 1857.

Resolved, That the Second Avenue Railroad Company be, and they are hereby, directed to have the rails between Forty-ninth and Sixty-first streets, in Second avenue, from the curb to the centre of the avenue, forthwith.*

Adopted by the Board of Aldermen, Oct. 20, 1859.

Adopted by the Board of Councilmen, Oct. 24, 1859.

---

* So printed in the proceedings of the two Boards. The resolution directed, or intended to direct, that the rails should be *removed* from the curb to the centre of the avenue.

Approved by the Mayor, Oct. 25, 1859.

Resolved, That the Second avenue, between Forty-eighth and Sixty-first streets, be paved with Belgian pavement, one-third the expense thereof to be paid by the owners of property, one-third by the Second Avenue Railroad Company, and one-third by the city, under the direction of the Croton Aqueduct Department, and that the accompanyng ordinance therefor be adopted.

Adopted by Board of Councilmen, Dec. 15, 1859.

Adopted by Board of Aldermen, Dec. 30, 1859.

Approved by the Mayor, Dec. 31, 1859.

THIRD AVENUE RAILROAD.

AGREEMENT, made the first day of January, in the year one thousand eight hundred and fifty-three, between the Mayor, Aldermen and Commonalty of the city of New York, parties of the first part, and the persons named in the resolutions hereinafter set forth, who shall duly sign and execute this agreement, their successors, associates and assigns, duly becoming parties thereto, as hereinafter provided, of the second part.

Whereas, The said parties of the first part, in Common Council convened, did, on the eighteenth day of December, one thousand eight hundred and fifty-two, duly pass and adopt the following resolutions, which were afterward, and on the thirty-first day of December, in the said year, returned by the Mayor without his approval or objections, and became operative and binding, in the words and figures following :

Resolved, That Myndert Van Schaick, Horace M. Dewey, John B. Dingledein, John Murphy, James W. Flynn, James McElvaney, Patrick McElroy, Thomas Murphy Philip Reynolds, Elijah F. Purdy, Bryant McCahill, George Caplin, Oscar F. Benjamin, and those who may hereafter become associated with them, have the authority and consent of the Common Council, and permission is hereby granted to them to lay a double track for a railroad in the following streets.

From a point at the intersection of Park row and Broadway, near the southwesterly corner of the Park, thence along Park row to Chatham street, thence along Chatham street to the Bowery, thence along the Bowery to the 3d avenue, and thence along the 3d avenue to the Harlem River, upon the following conditions, viz. :

Such track or tracks to be laid under the direction of the Street Commissioner, and on such grades as are now established, or may hereafter be established, by the Common Council, the said parties to become bound in a sufficient penalty to keep in good repair the space inside the tracks, and a space 2 feet each side of the same of each street in which the rails are laid ; and also that no steam power be used on any part of the road for propelling cars, and upon the further condition that said parties shall place new cars on said railroad, with all the modern improvements, for the convenience and comfort of passengers, and that they run cars thereon, each and every day, both ways and as often as the public convenience may require, under such prudential directions as the Common Council and the Street Commissioner may, from time to time, prescribe ; and

Provided, also, That the said parties shall in all respects comply with the directions of the Common Council in the building of the said railroad and in any other matter connected with the regulation of said railroad.

Provided, also, That the said parties shall, before this permission takes effect, enter into a good and sufficient agreement with the Mayor, Aldermen and Commonalty of the city of New York, to be drawn and approved of by the Counsel to the Corporation, binding themselves to abide by and perform the stipulations and provisions herein contained, and also all such other resolutions or ordinances as may be passed by the Common Council relating to the running of said cars over the said road.

And further, That they run a car thereon, each and every day, both ways, as often as every fifteen minutes from 5 to 6 A. M., every four minutes from 6 A. M. to 8 P. M., every fifteen minutes from 8 P. M. to 12 M., and as much oftener as public convenience may require, under such directions as the Common Council may, from time to time, prescribe.

Also, That the said passage on said railroad shall not exceed a greater sum than five cents for any distance, between the southern point of said railroad and Sixty-first street, and six cents for the entire length of said railroad.

And also, that said track or tracks shall be laid upon a good foundation, with a grooved rail, or such other rail as may be approved of by the Common Council and the Street Commissioner, even with the surface of the streets through which they may pass, and shall be commenced within six months, and completed to Forty-second street within one year from the passage of this resolution; and from Forty-second street, toward and to the Harlem River, as fast as the Third avenue shall be graded and in a proper condition to lay rails thereon.

2. *Resolved*, That said parties have the consent of the Common Council, and permission is hereby given to them, to connect their said railroad, at the junction of the Bowery and Grand street, with the Second Avenue Railroad, if constructed; and said parties, and those to whom permission may be given by the Common Council to lay a railroad through the Second avenue, shall have the free use in common of the double track from said junction through the Bowery to Chatham street, and of one of the tracks to be laid from the southerly termination of the Bowery through Chatham street to Pearl street; each of said parties to pay half the expense of constructing and keeping in repair the double and single track so to be used by them in common. Either of said parties to have the right to construct said double and single track so to be used in common, and, if constructed by either, the other of said parties shall pay half the actual cost thereof; or said parties may, by mutual agreement, construct the same jointly.

3. *Resolved*, That in consideration of the good and faithful performance of the conditions, stipulations, and agreement above prescribed, and of such other necessary re-

14

quirements as may hereafter be made by the Common
Council, for the regulations of the said railroad, the said
parties shall pay, from the date of opening the said railroad, the annual license fee for each car now allowed by
law, and shall have licenses accordingly.

4. *Resolved*, That within a reasonable time after the
passage of these resolutions, the said parties, or a majority
in interest thereof, may form themselves into an association
which shall be vested with all the rights and privileges
hereby granted ; and shall have power, by the votes of at
least a majority in interest of the associates, to frame and
establish articles of association providing for the construction, operation, and management of said railroad, and to
make contracts for the purchase of property for the use or
benefit of said railroad.

5. *Resolved*, That the association shall not be deemed
dissolved by the death or act of any associate, but his successor in interest shall stand in his place, and the rights of
each associate shall depend on his own fulfilment of the
conditions imposed on him by these restrictions or the articles of association and by-laws of the association ; and
in case of his failure to fulfill the same his rights shall be
forfeited to and devolve upon the remaining associates
after twenty days' notice of such failure, from the secretary of the association, specifying the particulars of his
delinquency; and the said parties or associates may, at any
time, incorporate themselves under the general railroad
act whenever two-thirds in interest of the associates shall
require it.

And whereas, It is deemed necessary by the said parties
of the first part, in order to preserve and duly effectuate

the grants, objects, stipulations, and intentions of the said resolutions, and for the purpose of more specifically determining the interest of the said parties in the rights and privileges granted by said resolutions, that provision should be made for an organization or association between the said parties of the second part, their successors, associates and assigns, duly admitted, according to this agreement, defining the mode in which the necessary capital for building the said railroad shall be contributed, and the manner in which the construction and management of the said railroad shall be conducted and controlled.

Now, it is hereby mutually declared, That the separate and individual interests of any or either of the said parties of the second part, their successors, associates, and assigns, in the said grant, and all licenses, rights and privileges, and powers conferred or provided for in said resolutions shall be conditioned and dependent upon the strict observance, performance and fulfilment by such persons of the terms of said resolutions and of this agreement ; and that, in case of failure to perform the same, and every part thereof, said grant shall be inoperative as to such person so failing, and his interest therein shall cease and determine ; said grant remaining operative in every respect as to all others of said parties, their successors, associates and assigns, and it is hereby covenanted, agreed and declared, by and between the parties aforesaid, viz. :

FIRST. The said parties of the second part, for themselves and their successors, associates and assigns, do hereby covenant and agree with the said parties of the first part, and with each other, that they will well and truly observe, perform, fulfill and keep the said resolutions hereinbefore particularly set forth, and all and every the provisions, stipulations, restrictions and conditions therein contained, and thereby imposed, according to the true intent and meaning thereof; it being understood that the rate of passage on said road shall not exceed five cents for any distance between the southerly point of said railroad and Sixty-first street, and six cents for the entire length of said railroad; and also that the said road shall be completed at the times and in the manner stated in said resolutions.

SECOND. The said parties of the second part, to the end that the provisions and intentions of the said resolutions may be carried into effect, the interests of the respective parties definitely ascertained, and the manner in which the construction and management of said road shall be conducted and controlled, effectually defined, do further covenant and agree with the said parties of the first part, and with each other, to associate and organize themselves together in the manner and upon the terms and conditions following, viz. :

Within ten days after this agreement is duly executed, the said parties of the second part, unless they, or a majority of them shall have previously organized themselves to the same effect as herein provided, shall and will organize themselves into an association or company, to be called the Third Avenue Railroad Company, for the purpose of constructing, operating and managing said railroad. The first meeting of the said parties to be called by the clerk of the Common Council, who shall, within three days after the due execution of this agreement, give, or cause to be given, a notice in writing, delivered to the per-

sons composing the said parties of the second part,
personally, or left at their residences or places of business,
specifying the time and place when and where such meeting shall be held.

The said parties of the second part, or as many of them
as shall meet, in pursuance of said order, shall thereupon
proceed, as before provided, to organize themselves into
the said Company, and shall have power and authority, by
the votes of the majority of the parties so assembled,

1. To estimate and declare the amount of capital requisite to construct the said railroad, provide cars, motive
power, stations, buildings, fixtures, and for all other expenses requisite to put the said railroad into thorough
practical operation.

2. To prescribe the mode in which the said capital and
all other sums that may thereafter be required for the
business of said Company shall be subscribed for, and the
time or times when the same shall be paid in, and the
manner in which the shares or interests of the parties
refusing or neglecting to subscribe or to pay may be forfeited.

3. To adopt suitable resolutions, by-laws, rules and
regulations for the organization of said Company, the subscription and payment of its capital, and all other sums
that may thereafter be required for its construction, operation and future business, the execution of contracts, the
liability of members, the terms, compensation, accountability, election, removal and duties of its officers, the disbursement of moneys, the transfer or assignment of shares
of its members and the entire management, direction and

control of its affairs, business, property and offices ; such by-laws may be altered, from time to time, in the manner prescribed therein.

4. The said parties of the second part shall be entitled to subscribe equally for the amount proposed as the original capital stock of said Company, and if any of them neglect to subscribe, or shall subscribe less than his proportion, the others may subscribe equally for the remainder, so as to make up a subscription for the whole amount. If for any reasons it shall be requisite to make other subscriptions, the persons who shall then be members of said Company shall be entitled to subscribe for the amount so required in proportion to the amounts of capital stock held by them, and if any shall neglect to subscribe, or shall subscribe for less than his proportion, the others may subscribe equally for the remainder.

5. Every person refusing or neglecting to subscribe to the capital stock of said Company, as originally declared, or to any subsequent increase thereof, or to pay his subscription, or any instalment thereof, at the times prescribed at the first meeting of said Company as aforesaid, or by the resolutions or by-laws of the said Company, all his rights, powers and privileges under said grant of the parties of the first part, and all his interests therein, shall be deemed to be freely and voluntarily waived and abandoned for the benefit of said Company and its remaining members, and shall cease, determine and be utterly null and void, and he shall be no longer a member of said Company, nor have any voice in the management of its affairs, nor any title and interest in its property : but such waiver and abandonment shall not be deemed to have taken

place until twenty days shall have elapsed after such person shall have had written notice of the required subscription or payment. But such person may, by a resolution, duly adopted by said Company, be reinstated in any or all of the rights, privileges and advantages so as aforesaid waived and lost, but upon such terms and conditions as may be hereby provided.

6. Every person who shall become a member of said Company shall thereby become a party to this agreement and all its conditions and stipulations, and the Company may direct the mode by which future members shall become so obligated, and no person shall become a member except on condition of becoming so obligated by agreement in writing duly executed.

7. The said railroad grant, property, rights and appurtenances, shall belong to, and be the property of, the persons who, for the time being, shall compose the said Third Avenue Railroad Company, in proportions equivalent to their shares of said capital stock, subject, however, to the management of the same in the manner herein provided.

8. Any shareholder may transfer his share or interest after he shall have paid one-third of his original subscription, on procuring the consent of a majority in interest of the shareholders, expressed by a resolution duly adopted, subject, however, to the provisions of this agreement, and on such terms and conditions as the by-laws may prescribe.

9. This Company shall not be dissolved by the death or insolvency of any of its members, nor by act or operation

of law, but in such and the like cases shall continue, and the persons becoming lawfully entitled to the shares shall become members of the said Company, and said Company shall have authority to incorporate themselves under the general railroad act, whenever two-thirds in interest of the shareholders shall require the same.

In witness whereof, to one of these presents, remaining with the said parties of the first part, the said parties of the second part have affixed their hands and seals ; and to the other part thereof, remaining with the said parties of the second part, the said parties of the first part have caused the common seal of the city of New York to be affixed, the day and year first above written.

M. VAN SCHAICK,	[L. S.]
H. M. DEWEY,	[L. S.]
JOHN B. DINGELDEIN,	[L. S.]
JOHN MURPHY,	[L. S.]
JAMES W. FLYNN,	[L. S.]
JAMES McELVANEY,	[L. S.]
PATRICK McELROY,	[L. S.]
THOMAS MURPHY,	[L. S.]
P. REYNOLDS,	[L. S.]
ELIJAH F. PURDY,	[L. S.]
BRYAN McCAHILL,	[L. S.]
GEORGE CALPIN,	[L. S.]
OSCAR F. BENJAMIN.	[L. S.]

Sealed and delivered in }
    presence of   }

HENRY E. DAVIES.

State of New York, City and County of New York, ss.:

On this first day of January, eighteen hundred and fifty-three, before me personally appeared Henry E. Davies, the subscribing witness to the above instrument, to me known, who, being by me duly sworn, did depose and say that he resides in the city and county of New York—that he knows Myndert Van Schaick, Horace M. Dewey, John B. Dingeldein, John Murphy, James W. Flynn, James McElvaney, Patrick McElroy, Thomas Murphy, Philip Reynolds, Elijah F. Purdy, Bryan McCahill, George Calpin, and Oscar F. Benjamin, the persons described in, and who executed the within instrument—that he was present, and saw them severally sign, seal and deliver the within instrument, for and as their act and deed ; and that they each severally acknowledged that they executed the same, and that thereupon he subscribed his name as a witness thereto.

J. MANSFIELD DAVIES,

Commissioner of Deeds.

———

Resolved, That the time within which, by the provisions of the grant dated January 1, 1853, authorizing the construction of the Third Avenue Railroad, the grantees in said grant named, and their assigns, were permitted to lay down a double track in the Bowery, south of Fifth street and along Park Row, be, and the same is hereby, extended until the expiration of three months after such time as the Third Avenue Railroad Company shall be deprived by the New York and Harlem Railroad Company of the privilege now enjoyed by the Third Avenue Railroad Company, of running their cars over the tracks of the New York and Harlem Railroad Company.

Adopted by the Board of Assistants, December 3, 1853.

Adopted by the Board of Aldermen, December 7, 1853.

Approved by the Mayor, December 9, 1853.

Resolved, That the Third Avenue Railroad Company be, and they are hereby, directed to cause Chatham street, from Pearl to Chambers street, to be repaired, according to the terms of their grant, within ten days from the adoption of this resolution.

Adopted by the Board of Councilmen, February 6, 1854.

Adopted by the Board of Aldermen, February 9, 1854.

Approved by the Mayor, February 14, 1854.

Petition of the Third Avenue Railread Co. for privilege to build a portico and balcony in front of their new depot. Prayer of the petitioner granted.

By the Board of Aldermen, June 8, 1857.

By the Board of Councilmen, June 11, 1857.

Approved by the Mayor, June 12, 1857.

The President of the Croton Aqueduct Board directed to notify the Third Avenue Railroad Co. to put in good repair, forthwith, all the pavements on and about their rails, and, in case of neglect or refusal, to have the same done at the expense of the Third Avenue Railroad Co.

Adopted by the Board of Aldermen, Sept. 10, 1857.

Adopted by the Board of Councilmen, Sept. 14, 1857.

Approved by the Mayor, Sept. 16, 1857.

Resolved, That the Third Avenue Railroad Co. be, and they are hereby permitted to substitute the iron instead of the Belgian pavement, on the up grades of Chatham street, at their own expense, under the supervision of the Croton Aqueduct Department.

Adopted by the Board of Aldermen, April 8, 1858.

Adopted by the Board of Councilmen, May 31, 1858.

Approved by the Mayor, June 9, 1858.

Resolved, That the Third avenue, from Fifty-sixth to Eighty-sixth street, be paved with the Belgian or trap-block pavement, the Third Avenue Railroad Company to pave or cause to be paved, at their own expense, all that portion of said avenue between and outside of their rails, which, by the terms of their grant from the city, they are required to keep in repair—the city at large, and the property owners on the line of said avenue, to pay an equal portion each of the expense of paving the remaining portion of the said Third avenue, between Fifty-sixth and Eighty-sixth streets, and that the accompanying ordinance be adopted.

Adopted by the Board of Aldermen, Oct. 10, 1859.

Adopted by the Board of Councilmen, Dec. 15, 1859.

Approved by the Mayor, Dec. 20, 1859.

NEW YORK AND HARLEM RAILROAD.

ARTICLES OF AGREEMENT made this ninth day of January, one thousand eight hundred and thirty-two, between the New York and Harlem Railroad Company, parties of the first part, and the Mayor, Aldermen and Commonalty of the city of New York, parties of the second part :

Whereas, an ordinance of the Common Council of the city of New York was passed by the Board of Aldermen on the sixteenth day of December last, and by the Board of Assistants on the nineteenth day of December last, and approved by the Mayor of the said city on the twenty-second day of December last, which ordinance is in the words and figures following, to wit :

A LAW *to authorize the New York and Harlem Railroad Co. to construct their Railway.*

SECTION 1. Be it ordained, &c., that the New York and Harlem Railroad Co. be, and they are hereby, permitted to construct and lay down, in pursuance of their act of incorporation, a double or single track, or railroad, or railway, along the Fourth avenue, from 23d street to the Harlem river, in conformity with a map now on file in the Register's office, and a branch thereof along 125th street, from the Fourth avenue to the Hudson river, provided that the width of such double railroad or way shall not exceed 24 feet.

§ 2. And be it further ordained, that if, at any time after the construction of the aforesaid railways by the said New York and Harlem Railroad Co., it shall appear to the Mayor,

Aldermen, and Commonalty of the city of New York, that the said railways, or any part thereof, shall constitute an obstruction or impediment to the future regulation of the city, or the ordinary use of any street or avenue (of which the said Mayor, Aldermen, and Commonalty shall be the sole judges), the said Railroad Company, or the directors thereof, shall, on the requisition of the said Mayor, Aldermen, and Commonalty, forthwith provide a remedy for the same, satisfactory to the said Mayor, Aldermen, and Commonalty; or, if they fail to find such remedy, they shall, within one month after such requisition, proceed to remove such railway, or obstruction or impediment, and to replace the street or avenue in as good condition as it was before the said railway was laid down; and, should the said Directors decline or neglect to obey such requisition, the said Mayor, Aldermen, and Commonalty may, upon the expiration of the time limited in such notice, cause the obstruction or impediment to be removed, and the avenues or streets restored as aforesaid, at the expense of the said Railroad Company.

§ 3. That the right of regulating the description of power to be used in propelling carriages on and along said railways, and the speed of the same, as well as all other power reserved to the said Mayor, Aldermen, and Commonalty, by the act of incorporation of the said Company, or any part thereof, be, and the same is hereby, expressly retained and reserved.

§ 4. That it shall .expressly be incumbent on the said New York and Harlem Railroad Co., at their own cost, to construct stone arches and bridges for all the cross streets now or hereafter to be made (which will be intersected by the embankments or excavations of the said railroad), and which, in the opinion of the Common Council, the public convenience requires to be arched or bridged, and also to make such embankments or excavations as (in the opinion of the Common Council) may be required to make the passage over the railroad and embankments, at the intersected cross streets, easy and convenient for all the purposes for which streets and roads are usually put to ; and also that the said Company shall make, at their own like cost and charges, all such drains and sewers as their embankments and excavations may (in the opinion of the Common Council) make necessary ; all which work to be done under the like requisition, and under like disabilities, as in the 2d section of this ordinance mentioned. And further, that the said Company shall make their railroad path, from time to time, conform to what may hereafter be the regulation of the avenue and road through which said railroad passes.

§ 5. That it shall be incumbent on the said New York and Harlem Railroad Company to commence and complete their said railroad in the respective times allowed for that purpose in their act of incorporation ; and, unless they commence and complete the same in the periods of time for the said commencement and completion in said incorporation specified, that then the consent of the Common Council, and all the powers and privileges given in this ordinance, shall cease and be null and void.

§ 6. That in case the said railroad should not be completed within the times for that purpose in their charter mentioned, or if, any time after the construction of the said railroad, the same should be discontinued, or not kept up and in repair as a good and sufficient railroad, that then the strip of land, to be taken for the said railroad, should be thrown open and become a part of the street or public avenue, without any assessment on the owners of the adjoining land or the public therefor.

§ 7. That no building shall be erected on the said strip of land to be taken for said railroad, and that a railing or other erections shall be made on the outer edges of the embankments or railroad path, and also such railing or fences on the edges of the excavations, as the Common Council shall, from time to time, deem necessary, to prevent accidents and loss of lives to our fellow-citizens.

§ S. That this ordinance shall not be considered as binding on the Common Council, nor shall the said ordinance go into effect, until the said Harlem Railroad Company shall first duly execute (under their corporate seal) such an instrument in writing (promising, covenanting, and engaging, on their part and behalf, to stand to, abide by and perform all the conditions and requirements in this ordinance contained) as the Mayor and the Counsel of the Board shall, by their certificate, approve, and not until such instrument shall be filed, so certified, in the Comptroller's office of this city.

Passed by the Board of Aldermen, December 16, 1831.

Passed by the Board of Assistants, December 19, 1831.

Approved by the Mayor, December 22, 1831

Now, this agreement witnesseth, that for and in consideration of the premises, and in pursuance of the requirements of the eighth section of said ordinance, the said parties of the first part do hereby for themselves, and their successors, promise, covenant and engage to and with the said parties of the second part and their successors and assigns, to stand, abide by and perform all the conditions and requirements in the said ordinance contained.

In witness whereof, the said parties of the first part have hereunto affixed their corporate seal, and caused the same to be signed by their Vice President (in the absence of their President), and attested by their Secretary, the day and year aforesaid.

JOHN MASON, [L. S.]
Vice President.

Witness,
  ISAAC ADRIANCE,
   Secretary pro tem.

Resolved, If the Board of Aldermen concur herein, that the maps presented by the New York and Harlem Railroad Company, so far as the same locate the route of the said railroad, from the north side of 23d street through the centre of the 4th avenue to Harlem River, and the branch of the same through the centre of 125th street, from the 4th avenue to the Hudson River, be approved, upon condition that neither this approval, nor anything herein contained, shall be construed into a consent to the said Company to construct the said railroad, but that the said Company shall first obtain the consent of the Mayor, Aldermen, and Commonalty of the City of New York before they commence the construction of said road.

Adopted by the Board of Assistant Aldermen, October 5, 1831.

Concurred in by the Board of Aldermen, October 10, 1831.

Approved by the Mayor, October 11, 1831.

Resolved, That the New York and Harlem Railroad Company be, and are hereby, authorized to take possession of the ground owned by the Common Council, over which the line of said railroad is ordered to be constructed, and that they be permitted to use the same during the continuance of the present charter for the purpose of a railroad and that only, and, when they cease so to use it, it shall revert to the Corporation ; provided always, that said land shall be so used as not to interfere with the use of the cross streets, and on condition, however, that if the said Corporation shall not commence the said railroad, and complete the same, within the time limited by their charter, then the privilege hereby granted shall cease and be void.

Adopted by both Boards, January 30, 1832.

Approved by the Mayor, February 1, 1832.

———

ARTICLES OF AGREEMENT, made this eighteenth day of May, one thousand eight hundred and thirty-two, between the New York and Harlem Railroad Company, parties of the first part, and the Mayor, Aldermen and Commonalty of the city of New York, parties of the second part.

Whereas, certain resolutions of the Common Council of the city of New York were passed by the Board of Aldermen on the second day of May, instant, by the Board of

15

Assistants on the seventh day of May, instant, and approved by the Mayor of the said city on the tenth day of May, instant, which resolutions are in the words and figures following, to wit:

Resolved, That the New York and Harlem Railroad Company be permitted, and the Common Council hereby consent, so far as their rights extend, that the said company may extend their rails southerly from the north line of 23d street to Prince street, subject, however, to the same conditions and restrictions which the Common Council heretofore imposed upon said Company in respect to that part of the road above 23d street. That the said Company may forthwith proceed to lay down a single track through the 4th avenue, south of 23d street, Union place, Bloomingdale Road, and Broadway, and another single track through the Bowery, both as far south as Prince street, and after two months' use of a single track upon the whole distance south of 23d street, on both Broadway and the Bowery, with convenient turnings at the several terminations above mentioned, they may, unless otherwise directed by the Common Council, lay down a second track on each of the above-mentioned routes, the same to be maintained by the said Company, subject at all times to the regulations of the Common Council, and also subject to the obligation of removing the whole or any part of the railways hereby permitted to be laid down, in case the Common Council shall hereafter see fit to require the same.

Provided, however, that all the said rails shall be laid down in such manner and in such parts of the said streets as shall be approved by the Street Commissioner, so as to cause no impediment to the common and ordinary use of the streets for all other purposes; and that the water courses of the streets shall be left free and unobstructed, and that the said Company shall pave the streets in and about the rails in a satisfactory and permanent manner, and keep the width of twenty feet of said paving, including the rails, in good repair at all times during the continuance of their use thereof.

And provided further, that if, at any time after the said rails shall have been laid down, the Common Council shall deem it necessary and shall order the said rails to be taken up, the said Railroad Company shall cause the pavement of the streets to be placed in good and sufficient repair.

And provided further, that the said Company have their single rail tracks above mentioned completed on or before the 1st day of May, 1834, and that they are to charge and receive such tolls, rates, or fare for the carrying of passengers or effects upon the said rail tracks, south of Twenty-third street, as the said Common Council may prescribe.

Resolved, That the above resolution shall not be considered as binding on the Common Council, nor shall the same go into effect until the said Harlem Railroad Company shall first duly execute, under their corporate seal, such an instrument in writing, promising, covenanting and agreeing, on their part and behalf, to stand to, abide by and perform all the conditions and provisions in the said resolution contained, as the Mayor and the Counsel of the Board shall approve of, by a certificate under their hands, nor until such instrument shall be filed, so certified, in the Comptroller's office of this city.

Now this agreement witnesseth, that for and in consideration of the premises, and in pursuance of the requirements of the said resolutions, the said parties of the first part do hereby, for themselves and their successors, promise, covenant and agree to and with the said parties of the second part, and their successors and assigns, to stand to, abide by, and perform all the conditions and requirements in the said resolutions contained.

In witness whereof, the said parties of the first part have hereunto affixed their corporate seal, and caused the same to be signed by their Vice President (in the abscence of their President), and attested by their Secretary, the day and year aforesaid.

JOHN MASON, [L. S.]

Vice President.

Attested,

 A. C. RAINETAUX,

  Secretary.

We hereby certify that we approve of the within, as being such an instrument in writing as the New York and Harlem Railroad Company are required to execute and file in the Comptroller's office, according to the second resolution above recited.

WALTER BOWNE, *Mayor.*

R. EMMET, *Counsel.*

NEW YORK, June 14, 1832.

Resolved, That the New York and Harlem Railroad Company be permitted, and the Common Council hereby consent, so far as their rights extend, that the New York and Harlem Railroad Company may continue their rails by single or double track southerly, from the north line of Prince street to the north line of Walker street, subject to the same conditions and restrictions which the Common Council heretofore imposed upon the said Company, in respect to that part of the said road between Prince street and Twenty-third street, as provided by the ordinances of the Common Council, May 20, 1832.

Adopted by the Board of Aldermen, April 24, 1837.

Adopted by the Board of Assistants, May 1, 1837.

Approved by the Mayor, May 4, 1837.

Resolved, That the New York and Harlem Railroad Company be permitted, and the Common Council hereby consent, that the said Company may continue their rails, similar to those laid down between Thirteenth and Fourteenth streets, by a double track from the Bowery through Broome street to Centre street, and from Broome street through Centre street to Chatham street, subject to the same conditions and restrictions which the Common Council heretofore imposed upon the said Company.

Resolved, That when such rails shall be laid through Centre and Broome streets, the said Company shall cause so much of the rails as are laid in the Bowery, south of Broome street, to be removed, and the street repaired, under the direction of the Street Commissioner.

Adopted by the Board of Aldermen, April 20, 1838.

Adopted by the Board of Assistants, May 2, 1838.

Approved by the Mayor, May 4, 1838.

Resolved, That the curb-stone on the easterly side of Centre street, in front of the market, be set six feet into the walk from its present line ; that the same be done. at the expense of the Harlem Railroad Company, and provided that the said Railroad Company pave the space between the rails and the said curb-stone with blocks of wood, the balance of the stone pavement to be allowed to their credit in the assessment for paving the street, or so much of the said pavement as does not legitimately belong to them to make.

Adopted by the Board of Aldermen, April 29, 1839.

Adopted by the Board of Assistants, April 29, 1839.

Approved by the Mayor, May 2, 1839.

Whereas, At the time the Common Council granted permission to the Harlem Railroad Co. to lay their rails in Centre street, it was understood that the said Company should pay for paving twenty feet in width through the centre of the said street ; and the said Company were assessed, and paid accordingly, for that part of Centre street between Grand and Walker streets.

And whereas, at the time the assessment was made for paving that portion of Centre street, between Walker and Chatham streets, it was believed that the said Company did not intend to avail themselves of the said privilege, and was not, therefore, assessed for the paving ; and whereas, the said Railroad Company have since recently laid their rails in the said street :

Therefore, Resolved, That the Street Commissioner request the Harlem Railroad Company to pay him the cost of paving twenty feet in width through the middle of Centre street, between Walker street and Chatham street, which has been assessed to the owners of property, and that he refund to the said owners their respective proportions of the same, when collected.

Adopted by the Board of Aldermen, July 5, 1839.

Adopted by the Board of Assistants, April 27, 1840.

Approved by the Mayor, April 30, 1840.

Resolved, That the tax of $836 25, of the New York and Harlem Railroad Company, in the Sixth Ward, for the year 1840, be remitted, and that the same be charged to errors and delinquencies of said Ward.

Adopted by the Board of Aldermen, March 15, 1841.

Adopted by the Board of Assistants, March 29, 1841.

Approved by the Mayor, April 1, 1841.

Resolved, That the Comptroller be authorized and directed to lease to the New York and Harlem Railroad Company the lot on Centre street, which they at present occupy, for one year from the expiration of their present lease, at a rent of $500, payable quarterly.

Adopted by the Board of Aldermen, January 16, 1843.

Adopted by the Board of Assistants, February 1, 1843.

Approved by the Mayor, February 8, 1843.

Resolved, That the Harlem Railroad Company are hereby required, on or before the first day of August next, to discontinue the use of steam power on the Fourth avenue south of the north line of Thirty-second street.

Adopted by the Board of Aldermen, Dec. 2, 1844.

Adopted by the Board of Assistants, Dec. 11, 1844.

Approved by the Mayor, Dec. 14, 1844.

Petition of the New York and Harlem Railroad Company, that they may be allowed three months to complete the necessary buildings at Thirty-second street, prior to the removal of their depot from Twenty-seventh street, was granted ; the said three months to commence from the date of the approval of the said petition by his Honor the Mayor.

By the Board of Aldermen, March 23, 1846.

By the Board of Assistants, March 23, 1846.

Approved by the Mayor, March 30, 1846.

Resolved, That the Corporation Attorney take legal measures to prevent the steam power of the Harlem Railroad Company from plying below Thirty-second street, on the Fourth avenue, as directed by the Mayor and Common Council in December, 1844.

Adopted by the Board of Aldermen, March 16, 1846.

Adopted by the Board of Assistants, March 16, 1846.

Received from his Honor the Mayor, March 30, 1846, without his approval or objections thereto ; therefore, under the provisions of the amended charter, the same became adopted.

Resolved, That the Harlem Railroad Company be required to construct a bridge of sufficient strength, and proper dimensions, for the transit of vehicles across the deep cut on the Fourth avenue, at each of the intersections of Thirty-fourth and Thirty-eighth streets, and that the same be erected without delay, and that the Clerk of the Common Council notify said Company thereof, on the adoption of this resolution.

Adopted by the Board of Assistants, November 2, 1846.

Adopted by the Board of Aldermen, November 9, 1846.

Approved by the Mayor, November 14, 1846

Resolved, That the New York and Harlem Railroad Company be, and they are hereby, directed, within thirty days from the passage of these resolutions, to restore the bridge crossing their track, at the intersection of Fiftieth street and Fourth avenue, in a firm and substantial manner.

Resolved, That said Company be, and they are hereby, directed to construct two bridges, one at the intersection of Seventy-ninth street and Fourth avenue, and one at the intersection of Eighty-fifth street and Fourth avenue ; said bridges to be built in a firm and substantial manner, and the same width as the bridge at Eighty-seventh street. Also, to inclose their track on Fourth avenue, with a fence or protection wall, along the edges, between Eighty-fourth street and the tunnel at or near Ninety-second street ; and also, that they inclose the sides of the bridge at the intersection of Eighty-seventh street and Fourth avenue.

Adopted by the Board of Assistants, January 24, 1848.

Adopted by the Board of Aldermen, February 1, 1848.

Approved by the Mayor, February 5, 1848.

Resolved, That the Harlem Railroad Company be requested to grade and regulate the 4th avenue from 28th to 32d street, the same now being in such an uneven condition as to render it impracticable to drive across the street, in consequence of raising the rails above the level of the street.

Adopted by the Board of Aldermen, April 26, 1848.

Adopted by the Board of Assistants, April 28, 1848.

Approved by the Mayor, May 1, 1848.

Resolved, That the New York and Harlem Railroad Company be authorized to lay down rails in Canal street, from their road in Centre street, to a point 75 feet east of Broadway, to enable them to afford increased accommodation for the public, which may be required by the extension of their own road, and by their connection with the New York and New Haven Railroad, and for the purpose of establishing a depot for passengers to and from New York and New Haven Railroad, with permission to cross the side-walk from the rail tracks into any premises which either of said Companies may become the lessees or owners of, all of which to be under the direction of the Street Commissioner, the privileges hereby granted to be enjoyed by said Company during the pleasure of the Common Council.

Adopted by the Board of Aldermen, Nov. 13, 1848.

Adopted by the Board of Assistants, Nov. 13, 1848.

Approved by the Mayor, Nov. 15, 1848.

Whereas, Resolutions having passed the Common Council on the 5th day of February, 1848, directing the Harlem Railroad Company to construct a bridge at the intersection of Fourth avenue and Eighty-fifth street, and to enclose their track with sufficient protection walls on said avenue, between Eighty-fourth and Ninety-second streets ; and

Whereas, Said Company have neglected to comply with the aforesaid directions, thereby endangering the lives of the inhabitants, (several having been killed) by falling in said road, caused by the said Company not complying with said resolutions ; therefore be it

Resolved, That the Counsel for the Corporation be, and is hereby authorized and directed to commence a suit against the Harlem Railroad Company, for the purpose of compelling said Company to erect a suitable bridge at the intersection of Fourth avenue and Eighty-fifth street, and also to enclose said road with a good and sufficient protection wall, from Eighty-fourth to Ninety-second street.

Adopted by the Board of Assistants, Feb. 18, 1850.

Adopted by the Board of Aldermen, March 11, 1850.

Approved by the Mayor, March 19, 1850.

Resolved, That the New York and Harlem Railroad Company be, and they are hereby, allowed to lay down (under the supervision of the Street Commissioner) a single track from their road in the Bowery, through 6th street to the rear of Tompkins market, between 6th and 7th streets, for the conveyance of country produce, &c., in order to improve the business of said market, and to accommodate that portion of our city, subject to all the restrictions contained in privileges to lay rails in **Centre** and other streets.

Adopted by the Board of Assistants, June 10, 1850.

Adopted by the Board of Aldermen, July 9, 1850.

Approved by the Acting Mayor, July 12, 1850.

Petition of residents of Harlem that permission may be granted to the New York and Harlem Railroad Company to lay a side track, or turnout, on the Fourth avenue, between 125th and 127th streets, was referred to the Street Commissioner, with power.

By the Board of Aldermen, July 9, 1850.

By the Board of Assistants, July 12, 1850.

Approved by the Acting Mayor, July 13, 1850.

Resolved, That the accompanying plan of curves, on the line of the New York and Harlem Railroad, at the intersection of Centre and Canal streets, Centre and Broome streets, and the Bowery, be referred to the Street Commissioner for his approval, with power to grant leave to the said Company to make said curves, if, in his opinion, the public interests would not be injured.

Adopted by the Board of Aldermen, July 11, 1850.

Adopted by the Board of Assistants, July 15, 1850.

Approved by the Mayor, July 18, 1850.

Resolved, That the New York and Harlem Railroad Company be, and they are hereby, directed to make, at their own cost, sustaining and parapet walls on the 4th avenue, **on** each side of their railroad track, from the northerly line of 32d street to the southerly line of 34th street, and also from the northerly line of 39th street to the southerly line of 42d street; also to make, at their own cost, proper sustaining walls along each side of their railroad track from the southerly line of 34th street to the northerly line of 39th street, and also, at their own cost, to build an arch over the said railroad track, between the said southerly line of 34th street and the northerly line of 39th street, and to build parapet walls across the 4th avenue, at the ends of the said arch, on the lines of 34th and 39th street ; the parapet walls to be 3 feet high, 2 feet thick at the base, and 1 foot 6 inches in thickness at the top, to be coped with cut granite, and surmounted with a neat iron railing, 2 feet in height, the sustaining walls to be of ordinary rubble masonry, 6 feet in thickness at the base, and 5 feet in thickness at the springing of the arch ; the arch to be of brick, 20 inches thick, with a radius and span of 24 feet, and to be of 15 in height, from the exterior rails of their rail track to the intrados of the arch, in conformity to a plan drawn by Edwin Smith, City Surveyor, dated May 15, 1850 ; also that the said New York and Harlem Railroad Company be directed, upon the completion of the said arch and walls, as above, to fill in and regulate that portion of the 4th avenue, between 34th and 39th streets, in conformity to the established grade line, and that the said work be commenced by the said Company on or before the 1st day of Oct., 1850, and that the same shall be completed by the said Company on or before the 1st day of May, 1851, under such directions as may be given by the Street Commissioner.

Adopted by the Board of Aldermen, July 12, 1850.

Adopted by the Board of Assistants, Aug. 6, 1850.

Approved by the Mayor, Aug. 8, 1850.

Resolved, That the New York and Harlem Railroad Company be, and they are hereby, authorized to take up their double track from the corner of Grand and Centre streets to the Bowery, and lay down a single track in the centre of the street, from the corner of Grand street, through Centre and Broome streets, to the Bowery, down the Bowery to Grand street, and through Grand street to Centre street, under the direction of the Street Commissioner.

Adopted by the Board of Aldermen, Sept. 11, 1850.

Adopted by the Board of Assistants, Sept. 12, 1850.

Approved by the Mayor, Sept. 13, 1850.

Petition of George Baker, Vice President of the Harlem Railroad Company, for permission to construct a sewer, or drain, from the corner of Tryon row to Chatham street, to connect with sewer in William street, at their own expense, under the direction of the Croton Aqueduct Department, was granted.

By the Board of Aldermen, October 7, 1850.

By the Board of Assistants, October 8, 1850.

Approved by the Mayor, October 11, 1850.

Resolved, That the New York and Harlem Railroad Company be, and they are hereby, authorized to lay grooved rails in a permanent manner, for a single track, on the westerly side of Chatham street, from the present terminus at Centre street, to the southerly end of the Park, with a turn-out, as shown on a profile on the petitions hereunto attached, for the exclusive use and purpose of running their City line of small passenger cars upon, to that point, subject to the pleasure and order of the Common Council; that when the Common Council may, or shall hereafter, order the said track to be taken up, the Company shall comply therewith at once, and that said Company shall, before said track is laid, execute to the City an agreement to comply therewith at once, when ordered, and that they will not run any but small passenger cars thereon, and that the Comptroller be charged with the preparation and execution of said agreement, said track to be laid under the direction of the Street Commissioner; provided that the said Company shall grade the street through which the said rails shall be laid, at their own expense, and keep the same in repair; that all ordinances heretofore pass'd, relative to the said Company, shall not be deemed to be in any way repealed by such permission hereby granted, except so far as the same conflicts therewith; and that said rails shall not be laid within a distance of 20 feet of the cross-walk at the corner of Broadway and the southern end of the Park.

Adopted by the Board of Assistants, Jan. 31, 1851.

Adopted by the Board of Aldermen, Feb. 4, 1851.

Approved by the Mayor, Feb. 6, 1851.

THIS AGREEMENT, made this sixth day of February, in the year one thousand eight hundred and fifty-one, between the Mayor, Aldermen and Commonalty of the

city of New York, of the first part, and the New York &
Harlem Railroad Company, of the second part, *Witnesseth*,
that the said parties of the first part, in consideration of the
sum of one dollar, to them paid by the said parties of the
second part, and of the premises and covenants herein contained, to be performed and kept by the said parties of the
second part, do hereby give, grant and convey, to the said
parties of the second part, the right and privilege of laying
grooved rails in a permanent manner for a single track, on
the westerly side of Chatham street, from the present terminus at Centre street to the southerly end of the Park,
with a turnout as shown on the profile hereto annexed ,
the privilege hereby granted being for the exclusive use
and purpose of the said parties of the second part, for running their city line of small passenger cars upon; the said
track to be laid under the direction of the Street Commissioner of the city of New York

It is further understood, by and between the parties to
these presents, that the said rails shall not be laid within a
distance of twenty feet of the crosswalk at the corner of
Broadway and the southern end of the Park

And the said parties of the second part, for and in consideration of the premises, do hereby covenant and agree,
to and with the said parties of the first part and their
successors, in manner following, namely : that the privilege
and consent hereby granted to the said parties of the second
part are subject to the pleasure and order of the Common
Council of the said parties of the first part ; and that when
the said Common Council shall hereafter order or direct the
said track, so permitted by these presents to be laid, to be

taken up and removed, that the said parties of the second part and their successors, will comply with such order or direction forthwith, and remove said track and rails, and replace and restore that part of said street upon which the same were laid, in the same plight and condition as the same was in before the laying of said track or rails, at the expense of the said parties of the second part and their successors.

And the said parties of the second part do further covenant and agree that they will grade all that portion of the said street occupied by said track and four feet on each side thereof, and keep the same in good repair, at their own expense, so long as the same shall be occupied and used by the said parties of the second part, or their successors, for the purposes aforesaid.

And the said parties of the second part do further covenant and agree to fulfill and comply with all the stipula tions and agreements herein contained, on their part to be performed and fulfilled.

And it is further mutually understood and agreed, by and between the parties to these presents, that, in case of the non-compliance by the said parties of the second part, or their successors, of any of the stipulations, covenants or agreements herein contained, on their part to be performed, fulfilled and kept, that then and from thenceforth the privilege herein granted to the said parties of the second part shall cease and determine.

In witness whereof, to one part of these presents, remaining with the said parties of the second part, the said parties

16

of the first part have caused the common seal of the city
of New York to be affixed; and to the other part thereof,
remaining with the said parties of the first part, the said
parties of the second part have caused their common seal
to be affixed, the day and year first above written.

The New York and Harlem Railroad Company, by

ROBERT SCHUYLER, [L. S.]

President.

CITY AND COUNTY OF NEW YORK, ss. ·

On this 27th day of February, 1851, before me personally
appeared Robert Schuyler, who is personally known to me,
who, being by me duly sworn, did depose and say that
he is the President of the New York and Harlem Railroad
Company; that the seal annexed to the foregoing instrument is the seal of the said company, and was affixed
thereto by authority of said company; and that he, the
deponent, resides in the Fifth Ward of the city of New
York.

CHARLES W. SANDFORD,

Commissioner of Deeds, New York.

———

Resolved, That the New York and Harlem Railroad Company be permitted to construct a branch or side track from
the tracks in the 4th avenue, to a point in front of their
depot on 26th street, distant 300 feet westerly from the
northwesterly corner of 26th street and 4th avenue, with
a single line of rails, and that such permission be granted

for a period of six months from the 1st day of January, 1851, and that the rails be taken up at the expiration of the period asked for by the Company, under the direction of the Street Commissioner.

Adopted by the Board of Aldermen, Feb. 5, 1851.

Adopted by the Board of Assistants, June 3, 1851.

Approved by the Mayor, June 4, 1851.

Resolved, That the New York and Harlem Railroad Company be, and they are hereby, directed to erect, without delay, bridges across their road at 83d, 84th, and 88th streets, the same as that erected across said road at 85th street.

Adopted by the Board of Assistants, May 30, 1851.

Adopted by the Board of Aldermen, June 4, 1851.

Approved by the Mayor, June 13, 1851.

Resolved, That the Street Commissioner be, and he is hereby, directed to have the railroad track, extending from the Harlem Railroad, in Centre street, up Chambers street toward Broadway, taken up within ten days from the passage of this resolution.

Adopted by the Board of Assistants, May 6, 1851.

Adopted by the Board of Aldermen, June 4, 1851.

Approved by the Mayor, June 13, 1851.

Resolved, That a space forty feet in width, and extending through the middle of the Fourth avenue, from Thirty-fourth to Thirty-eighth street, be, and the same is hereby, appropriated for the purpose of a public park or pleasure-ground; and that the same be laid out, under the direction of the Street Commissioner, in accordance with the general plan herewith accompanied. And that the same be carried into effect, as soon as the Harlem Railroad Company shall have completed the arching of the Fourth avenue, and that the acompanying ordinance be adopted therefrom.

Adopted by the Board of Aldermen, August 19th, 1851.

Adopted by the Board of Assistants, September 9th, 1851.

Approved by the Mayor, October 7th, 1851.

Resolved, That the New York and Harlem Railroad Company be permitted to reduce the grade of the Fourth avenue, on the east side of the road, between Thirty-second and Thirty-fourth streets, to a level with their track, with a turn-out on the east side thereof, entering the block at a point north of the north side of Thirty-third street, in accordance with the accompanying diagram, on condition that they, at their own expense, widen said avenue twenty-five feet, on the west side, between Thirty-second and Thirty-fourth streets, and cause the land for this additional width to be ceded to the Corporation of the city of New York, as part of the said Fourth avenue.

Adopted by the Board of Assistants, December 10, 1851.

Adopted by the Board of Aldermen, December 26, 1851.

Approved by the Mayor, December 29, 1851.

Resolved, That the New York and Harlem Railroad Company be directed to have the iron railing, on the bridge at the crossing of Thirty-fourth street and Fourth avenue, secured in a proper manner immediately.

Adopted by the Board of Aldermen, December 16, 1851.

Adopted by the Board of Assistants, December 29, 1851.

Approved by the Mayor, December 30, 1851.

Resolved, That the New York and Harlem Railroad Company be, and they are hereby, directed to construct a new and substantial bridge for their road at 104th street immediately, under the direction of the Street Commissioner; and further, if said Company fail to comply herewith, then the Street Commissioner is hereby directed to erect the same at the expense of said New York and Harlem Railroad Company, without delay.

Adopted by the Board of Assistants, September 18, 1852.

Adopted by the Board of Aldermen, September 20, 1852.

Approved by the Mayor, September 21, 1852.

Resolved, That the New York and Harlem Railroad Company be, and they are hereby, directed to take up the rails of their track in Canal street, and relay the same with the *grooved* rail, in the same manner as that laid along the east side of the Park; said track in Canal street not to extend west of the depot near Broadway, and that the same be completed within sixty days from the passage of this resolution.

Adopted by the Board of Aldermen, October 11, 1852.

Adopted by the Board of Assistants, November 9, 1852.

Approved by the Mayor, November 10, 1852.

Resolved, That the Harlem and New Haven Railroad Company shall station a man on the northwest corner of Grand street and the Bowery, to warn persons coming down the Bowery, on foot or in vehicles, of the near approach of the cars toward the corner of Grand street and the Bowery.

Adopted by the Board of Assistants, January 14, 1853.

Adopted by the Board of Aldermen, February 9, 1853.

Approved by the Mayor, February 15, 1853.

Resolved, That the Harlem Railroad Company be, and is hereby, authorized to lay a track in connection with their railroad at Hamilton square for the accommodation of the New York State Agricultural Society, during the period of their approaching fair, the same to be removed immediately after the fair is closed.

Adopted by the Board of Councilmen, September 4, 1854.

Adopted by the Board of Aldermen, September 18, 1854.

Approved by the Mayor, September 19, 1854.

Resolved, That the New York and Harlem Railroad

Company be, and are hereby, directed to build a good substantial wooden bridge, the full width of the street, similar to the one at 86th street, across the cut at 90th street, and the same to be done under the direction of the Street Commissioner.

Adopted by the Board of Aldermen, July 6, 1854.

Adopted by the Board of Councilmen, December 19, 1854.

Approved by the Mayor, December 20, 1854.

Resolved, That no locomotive or steam engine be allowed to run on the tracks of the Harlem or New Haven Railroad Company, on 4th avenue, south of 42d street, 18 months after the passage of this ordinance.

Adopted by the Board of Aldermen, December 7, 1854.

Adopted by the Board of Councilmen, December 22, 1854.

Approved by the Mayor, December 27, 1854.

Resolved, That the Harlem Railroad Company be directed to station a flagman at the corner of the Bowery and Broome street, for the purpose of warning pedestrians and those persons driving vehicles of the nigh approach of the rail cars as they turn the corner of the said Bowery and Broome street.

Adopted by the Board of Aldermen, January 22, 1857.

Adopted by the Board of Councilmen, February 2, 1857.

Approved by the Mayor, February 3, 1857.

Resolved, That the Harlem Railroad Company be directed to place a flagman at the corner of Pearl and Centre streets, for the protection of persons crossing said streets.

Adopted by the Board of Councilmen, February 6, 1857.

Adopted by the Board of Aldermen, February 9, 1857.

Approved by the Mayor, February 10, 1857.

Resolved, That the Street Commissioner be, and he is hereby, instructed to cause the Harlem Railroad Company to put their track in thorough condition forthwith, and if the said Harlem Railroad Company fail to comply within one week after the passage of this resolution, that then the 3d Avenue Railroad Company have permission, and the permission is hereby granted to them, to lay a new track on the line of the Bowery, under the direction of the Street Commissioner, from 5th street to Grand street, instead of the old T rail ; and the said 3d Avenue Railroad Company shall enjoy all the privileges now held by the Harlem Railroad Company below 5th street, on the line of the Bowery ; and all permissions and privileges heretofore granted to the Harlem Railroad Company, conflicting with the above, be, and the same are hereby, annulled and repealed.

Adopted by the Board of Aldermen, Feb. 16, 1857.

Adopted by the Board of Councilmen, Feb. 18, 1857.

Approved by the Mayor, Feb. 21, 1857.

The President of the Croton Aqueduct Board directed to notify the Harlem Railroad Company to put in good repair, forthwith, all the pavements in and about their rails, and, in case of neglect or refusal, to have the same done at the expense of the Railroad Company.

Adopted by the Board of Aldermen, Sept. 10, 1857.

Adopted by the Board of Councilmen, Sept. 14, 1857.

Approved by the Mayor, Sept. 16, 1857.

Whereas, The Harlem Railroad Company having failed to coerce the Common Council in repealing the existing ordinances of the Corporation, requiring them to discontinue the use of steam below 42d street; and

Whereas, In consequence of the said failure, the said Company having determined to defy the acts of the Common Council, and positively refuse to obey the ordinances of this body, and are now, through their paid agents, endeavoring to secure the passage of an act from the Legislature of the State (a bill having been introduced for this purpose), to continue their present nuisance, in violation of the existing ordinances, and their agreement, made with the Corporation ; be it therefore

Resolved, That the Counsel to the Corporation be, and he is hereby, instructed to prepare a remonstrance (in behalf of the Common Council), against the passage of any bill giving to the Harlem Railroad Company the privilege to continue the running of locomotives and the use of steam below 42d street, in violation of the ordinances of the Mayor and Commonalty of this city.

Adopted by the Board of Aldermen, Feb. 18, 1858

Adopted by the Board of Councilmen, Feb. 23, 1858.

Approved by the Mayor, Feb. 24, 1858.

Resolved, That the New York and Harlem Railroad Company be, and they are hereby, directed to take up the pavement and rail tracks in Broome street, from a point 100 feet west of Mulberry street to the easterly side of Elizabeth street, and relay the same in accordance with the original grade of that part of said street, within 10 days from the date of the passage of this resolution, and that, in case of refusal or neglect, on the part of said Company, to comply with this resolution, that the Croton Aqueduct Board be, and they are hereby, directed to remove said tracks from said part of Broome street, and to restore the carriage way thereof to the aforesaid grade without delay.

Adopted by the Board of Aldermen, March 31, 1858.

Adopted by the Board of Councilmen, April 5, 1858.

Approved by the Mayor, April 13, 1858.

Resolved, That the Harlem Railroad Company be, and they are hereby, directed to cause their small cars to be run on their track to 42d street as often and as regularly as they are now run between 27th street and Park row, the said Company to commence running said cars, as aforesaid, within four months after the approval of this resolution by his Honor the Mayor.

Adopted by the Board of Councilmen, May 10, 1858.

Adopted by the Board of Aldermen, June 21, 1858

Approved by the Mayor, July 12, 1858.

AN ORDINANCE

In relation to the New York and Harlem Railroad Company.

Be it ordained by the Mayor, Aldermen, and Commonalty of the city of New York, in Common Council convened :

SECTION. 1. The New York and Harlem Railroad Company is hereby authorized, empowered and permitted to use steam in the drawing of their passenger and freight cars upon their railroad on the Fourth avenue, to and from the northern extremity of Manhattan, or New York Island, to the south side of 42d street, and to permit the use thereof by the New York and New Haven Railroad Company to the same point, with turn-outs to the engine-houses respectively, for a period of thirty years, from the passage of this ordinance.

§ 2. Until the completion of their new machine shops at or above 42d street, the New York and Harlem Railroad Company shall be permitted to run their engines with steam, for repairs only, but without any car, truck or other vehicle attached, to and from their present machine shop at 32d street ; but such permission shall not extend, in any event, beyond a period of eighteen months from the date of this ordinance

§ 3. The New York and Harlem Railroad Company are hereby authorized to lay down a double track or railway from their track in the Fourth avenue, at 42d street, up said street to Madison avenue, and up Madison avenue to

79th street, or as far as it may, from time to time, be opened, for the use of their small cars only.

§ 4. The said Company is hereby authorized to lay down in the Fourth avenue, between 42d and 50th streets, two additional tracks, for the use of themselves and the New York and New Haven Railroad Company, to enable them to land and receive their passengers, and may cover that portion thereof which extends from 42d to 44th street by a neat ornamental roof or shed, to be first approved by the Mayor of the city of New York, and that the sidewalks opposite to said building be reduced to sixteen feet on each side of said avenue, in front of the premises of said Railroad Company.

§ 5. The New York and Harlem Railroad Company shall forthwith complete the title of the Corporation of the city of New York to the strip of ground, 20 feet wide, between 33d and 34th streets, on the west side of the Fourth avenue, and also to the strip of ground, 20 feet wide, between 32d and 33d streets, agreed by them to be conveyed to the city, and shall, within six months from this date, remove their engine-house at 32d street from said last-mentioned strip of land.

§ 6. In the case the New York and Harlem Railroad Company shall fail to carry out in good faith the provisions of the second and fifth sections of this ordinance, within the times in said sections respectively limited, the privileges hereby granted shall cease and determine, and the ordinance shall be null and void.

Adopted by the Board of Aldermen, Dec. 21, 1858.

Adopted by the Board of Councilmen, Dec. 22, 1858.

Approved by the Mayor, Dec. 31, 1858.

Resolved, That Fourth avenue, from Thirty-fourth to Thirty-eighth street, be paved with trap-block or Belgian pavement, at the expense of the owners of property benefited thereby, and that the Harlem Railroad Company be compelled to reduce the grade of the Fourth avenue, at its intersection with Thirty-fourth street, in accordance with the annexed ordinance, and that the accompanying ordinance therefor be adopted.

Adopted by the Board of Councilmen, Sept. 29, 1859.

Adopted by the Board of Aldermen, Oct. 3, 1859.

Approved by the Mayor, Oct. 4, 1859.

Resolved, That the New York and Harlem Railroad Company be, and they are hereby, directed to construct a substantial bridge over their road at Seventieth street, immediately, under the direction of the Street Commissioner ; the said bridge to be similar to the bridge now at Eighty-seventh street; and further, if said company fail to comply with the conditions of this resolution, then the said Street Commissioner is hereby authorized and directed to erect, or cause to be erected, the said bridge over Seventieth street, at Fourth avenue, at the expense of the said New York and Harlem Railroad Company, without delay.

Adopted by the Board of Aldermen, Oct. 24, 1859.

Adopted by the Board of Councilmen, Oct. 27, 1859.

Approved by the Mayor, Oct. 29, 1859.

NEW YORK AND NEW HAVEN RAILROAD.

Resolved, That the block of ground bounded by Centre, Franklin, Elm and White streets, be leased to the New York and New Haven Railroad Company, for the term of twenty-one years (the Arsenal buildings, on Elm and Franklin streets, to remain for the use of the military until May 1st, 1851), at an annual rent of six thousand dollars, payable quarterly, together with the taxes and assessments on the same, with covenants for renewal at the expiration of twenty-one years, at a rent to be fixed by appraisement. The said company to improve the said premises within one year from the date of their lease. Said premises to continue during said lease for a railroad depot, and the Comptroller is hereby directed to have a lease executed in accordance with this resolution, provided that nothing therein contained shall be construed or taken as a consent or assent, on the part of the Corporation of the city of New York, to the use by the said railroad company of any of the streets and avenues of said city, for the purpose of running cars thereon, by virtue of an agreement with the Harlem Railroad Company, or as a waiver of the right and power of the Common Council of said city to regulate and control the said New York and New Haven Railroad Company, to the same extent it can now control the New York and Harlem Railroad Company.   .

Adopted by the Board of Aldermen, September 3, 1850.

Adopted by the Board of Assistants, September 4, 1850.

Approved by the Mayor, September 5, 1850.

THIS INDENTURE, made the fifth day of September, in

the year of our Lord one thousand eight hundred and fifty, between the Mayor, Aldermen, and Commonalty of the city of New York, of the one part, and the New York and New Haven Railroad Company, of the other part—

Witnesseth, that the said Mayor, Aldermen, and Commonalty of the city of New York, for and in consideration of the rents, covenants, payments, articles and agreements, hereinafter mentioned and contained on the part of the said the New York and New Haven Railroad Company, their successors and assigns, to be paid, done, performed, fulfilled and kept, have demised and to farm letten, and by these presents do demise and to farm let, unto the said the New York and New Haven Railroad Company, ALL the certain block of ground situate, lying and being in the Sixth Ward of the city of New York, and bounded, described and containing as follows: that is to say—northerly by White street; easterly by Centre street; southerly by Franklin street; and westerly by Elm street, as laid down on a map hereto annexed, drawn by John J. Sewell, City Surveyor, dated New York, September 13, 1850, said map being considered a part of this Indenture, and reference thereto had.

To HAVE AND TO HOLD the said above-described premises unto the said the New York and New Haven Railroad Company, their successors and assigns, from the first day of September, Anno Domini one thousand eight hundred and fifty, for and during the full end and term of twenty-one years from thence next ensuing, and fully to be complete and ended, yielding and paying therefor, yearly, and every year during the said term, unto the said Mayor, Aldermen and Commonalty of the city of New York, their successors and assigns, the yearly rent of six thousand dollars, lawful money of the United States of America, which said yearly rent shall be paid on the usual quarterly days of payment; that is to say, one thousand five hundred dollars on the 1st day of August, and the like sum on the first days of November, February and May in each and every year during the said term, the first payment to be made on the first day of August next.

Provided always, That if it should so happen that the said yearly rent of 6,000 dollars, or any part thereof, shall be behind or unpaid, by the space of 10 days after any day of payment, on which the same ought to be paid as aforesaid, or if the said party of the second part, their successors or assigns, shall neglect or omit to pay, do, perform, fulfill or keep any or either of the payments, covenants, articles, clauses, agreements, matters or things herein contained, which on the part or behalf of the said party of the second part, their successors or assigns, are to be paid, done, performed, fulfilled or kept during the term aforesaid, according to the true intent and meaning of these presents, that then (and in every such case or cases), and at all times thereafter, it shall and may be lawful to and for the said Mayor, Aldermen and Commonalty of the city of New York, their successors and assigns, into all the said demised premises, and every part thereof, wholly to re-enter, re-possess, and to have and to enjoy the same again as in their former estate; and the said The New York and New Haven Railroad Company, their successors or assigns thereout, and from thence to expel and remove, anything herein contained to the contrary notwithstanding. And the said The New York and New Haven Railroad Company, for themselves, their successors and assigns, doth covenant and grant to and with the said Mayor, Aldermen and Commonalty of the city of New York, their successors and assigns, by these presents, that they, the said The New York and New Haven Railroad Company, their successors and assigns, shall and will yearly, and every year during the term hereby demised, well and truly pay unto the said Mayor, Aldermen and Commonalty of the city of New York, their successors and assigns, the yearly rent of six thousand dollars, on the days and times hereinbefore limited for the payment thereof, without fraud or delay, and that they, the said The New York and New Haven Railroad Company, their successors and assigns, shall and will, at their own proper costs and charges, bear, pay and discharge all such duties, taxes, assessments, impositions and payments, extraordinary as well as ordinary, as shall, during the term hereby demised be issued or grow due and payable out of, and for the said demised premises, or which shall during the said term be laid, assessed or imposed upon the said premises, or upon the owners or occupants thereof, for and in respect to the same, by virtue of any existing or future law of the United States of America, or of any existing or future law of the Legislature of the State of New York, or of any existing or future law or ordinance of the Mayor, Aldermen and Commonalty of the City of New York : to the end that the said yearly rent hereby reserved shall, during the term demised, be received by the said Mayor, Aldermen and Commonalty, and their successors and assigns, free and clear from any deduction ; and that they shall be at no expense, cost or charge whatsoever, for or in respect to the said demised premises, during the said term.

17

AND the said the Mayor, Aldermen and Commonalty of the City of New York, for themselves, their successors and assigns, do covenant and agree to and with the said The New York and New Haven Railroad Company, their successors and assigns, that at the expiration of the term hereby demised, they, the said the Mayor, Aldermen and Commonalty of the City of New York, will execute to the said The New York and New Haven Railroad Company a renewal of this lease for a term of twenty-one years, from the expiration of this term, with like covenants as are contained in this lease except the covenant for renewal; the rent to be paid under such lease to be ascertained and determined by two sworn appraisers, to be chosen one by the said Mayor, Aldermen and Commonalty, and one by the said Railroad Company, or by a sworn umpire to be chosen by the said appraisers in case they cannot agree upon the said rent to be paid.

AND the said The New York and New Haven Railroad Company do hereby covenant and agree to and with the said the Mayor, Aldermen and Commonalty of the city of New York, that they will, within one year from the first day of May, 1851, improve the said premises hereby demised for a railroad depot, and that they will use the said premises hereby demised for and as a railroad depot during the term of this lease.

AND it is mutually understood and agreed, by and between the parties to this lease, and it is upon the express condition that nothing herein contained shall be construed or taken as a consent, or assent, on the part of the said the Mayor, Aldermen and Commonalty of the city of New York, to the use by the said The New York and New Haven

Railroad Company, of any of the streets, avenues of the said city, for the purposes of running cars thereon by virtue of any agreement with the Harlem Railroad Company, or as a waiver of the right and power of the Common Council of said city to regulate and control The New York and New Haven Railroad Company to the same extent as it can now control The New York and Harlem Railroad Company.

AND it is further agreed, that the Arsenal Buildings, on Elm and Franklin streets, shall remain for the use of the military until May 1st, A. D. 1851, and the said parties of the second part agree to permit the same to be used for that purpose for the time aforesaid.

And the said The New York and New Haven Railroad Company, for themselves, their successors and assigns, do further covenant and grant, to and with the Mayor, Aldermen and Commonalty of the city of New York, their successors and assigns, that the said The New York and New Haven Railroad Company, their successors and assigns, shall and will, well and truly, on the last day of the said term, hereby demised, or other sooner determination thereof, deliver up the said hereby demised premises, into the hands and possession of the said Mayor, Aldermen, and Commonalty of the city of New York, their successors or assigns, without fraud or delay, unless the said lease shall be renewed in the manner hereinbefore mentioned: And, lastly, That if the said The New York and New Haven Railroad Company, their successors or assigns, shall and do assign, or make over, all or any part of the premises hereby demised, to any person or persons whatsoever, without the leave and approbation of the said Mayor, Aldermen and Commonalty of the city of New York first had and obtained, then and in that case this Indenture, and everything herein contained, shall cease, determine and be utterly null and void, anything herein contained to the contrary thereof notwithstanding.

In witness whereof, to one part of these presents, remaining with the Mayor, Aldermen and Commonalty of the city of New York, the said The New York and New Haven Railroad Company have put their corporate seal, attested by the signatures of their President and Secretary; and to the other part thereof, remaining with the said The New York and New Haven Railroad Company, the said Mayor, Aldermen and Commonalty of the city of New York have caused the seal of the city of New York to be affixed, the day and year first above written.

The New York and New Haven   [L. S.]
    Railroad Company, by

              ROBERT SCHUYLER,

                        President, &c., &c.

WILLIAM P. BURRALL,

Secretary New York and New Haven Railroad Company.

CITY AND COUNTY OF NEW YORK, ss. : On the twenty-seventh day of November, one thousand eight hundred and fifty, before me came Robert Schuyler, who is personally known to me to be the President of The New York and New Haven Railroad Company, the Corporation named in the foregoing instrument, and the said Robert Schuyler, being by me duly sworn, did depose and say that he resides in the city of New York, that the seal affixed to the said instrument is the corporate seal of the said Company, and was affixed thereto by their authority.

JOSEPH STRONG,

Commissioner of Deeds.

———

BROADWAY RAILROAD.

Resolved, That Jacob Sharp, Freeman Campbell, Wm. B. Reynolds, James Gaunt, J. Newton Squire, Wm. A. Mead, David Woods, John L. O'Sullivan, Wm. M. Pullis, Jonathan Roe, John W. Hawkes, James W. Faulkner, Henry Du Bois, John J. Hollister, Preston Sheldon, John Anderson, John R. Flanagan, Sargent V. Bagley, Peter B. Sweeny, Charles B. White, James W. Foshay, Robert E. Ring, Thomas Ladd, Conklin Sharp, Samuel L. Titus, Alfred Martin, D. R. Martin, William Menzies, Charles H. Glover, Gershon Cohen, and those who may, for the time being, be associated with them, all of whom are herein designated as associates of the Broadway Railway, have the authority and consent of the Common Council to lay a double track for a railway in Broadway and Whitehall, or State street, from the South Ferry to 59th street, and also, hereafter, to continue the same, from time to time, along the Blooming-dale road, to Manhattanville, which continuation they shall be required, from time to time, to make, whenever directed by the Common Council, the said grant of permission and authority being upon and with the following conditions and stipulations, to wit :

First. Such tracks shall be laid under the direction of the Street Commissioner, in, or near the middle of the street, the outer rails not exceeding 12 feet 6 inches apart, and the rails being laid flush and even with the pavement, the inner portion of the rail being of equal height with the outer, with grooves not exceeding 1 inch in width, or such other rails as shall be approved by the Street Commissioner, or the Common Council, on such grades as are now established, or may hereafter be established by the Common Council; and the said associates shall keep in good repair the space between the said rails, and one foot on each side, and no motive power, excepting horses, shall be used below 59th street.

Second. The said associates shall place new cars on said railroad, with all the modern improvements for the convenience and comfort of passengers ; and they shall run cars thereon every day, both ways, as often as the public convenience may require, under such direction as the Common Counsel may, from time to time, prescribe ; said cars, with the horses attached, not to exceed 45 feet in length.

Third. The said associates shall, in all respects, comply with the directions of the Common Council, in the building of such railway, and in the running of the cars thereon.

Fourth. At the Bowling Green, the said associates may divide the two tracks aforesaid, running one of them down Whitehall street, and the other down State street, should they deem such division necessary ; and also whenever, in the course of their route, the said road shall pass a public square, it may be carried, with a single track, round both sides of said square instead of only one, for the better accommodation of the public on both sides thereof.

Fifth. The said associates shall be required to procure a depot at some place near or at the lower part of said route, for the purpose of keeping withdrawn from Broadway such proportion of the cars coming down in the morning as shall not be required for the accommodation of the return travel, until the afternoon ; and also, they shall be required to stop a portion of the cars at the Park, and to send down, below that point, no greater proportion of the whole number employed than shall be found, by experience, to be requisite for the accommodation of the travel below that point, subject to regulation by the Common Council.

Sixth. The cars shall be so constructed as not to make provision intended for standing passengers to crowd upon the seated passengers, and also, when all the seats are full, the cars shall not be stopped to take in more passengers to be crowded into the said seats, a flag being displayed in front of the car to give notice that all the seats are full.

Seventh. The said cars shall not be allowed to stop, so as to obstruct a crossing, nor to stop more frequently in a block (unless the same be of extraordinary length), than just beyond its first crossing, except in rainy weather.

EIGHTH. The said associates shall keep an attendant, distinguishable by some conspicuous mark or badge, at every such appointed stopping place, in all the parts of the street usually much crowded with vehicles, whose duty it shall be, with attention and respect, to help in and out of the cars all passengers who may desire such assistance, and in general to watch over the safety of passengers from all dangers of passing vehicles.

NINTH. The said associates shall be required to keep, or cause to be kept in readiness, a number of sleighs adequate to the public accommodation, when the travel of the cars may be obstructed by snow.

TENTH. The said associates shall cause the said street to be well swept and cleaned every morning, and sweepings carried away, before 8 o'clock in summer and 9 o'clock in winter, except Sundays, this provision applying to the whole of the street south of Fourteenth street, above which point the same shall be done as often as twice a week, when the weather will permit.

ELEVENTH. No higher rate of fare shall be charged, for the conveyance of passengers from any one point to any other point along said route, and such combined systems of routes as may hereafter be adopted by means of cars and transient omnibuses, than five cents for each passenger.

TWELFTH. In consideration of the good and faithful performance of all these conditions, stipulations, and requirements, and of such other requirements as may hereafter be made by the Common Council, for the regulation of the said, railway, as aforesaid, the said associates shall pay, for 10 years from the date of opening the said railway, the annual license fee, for each car, now allowed by law, and shall have a license accordingly; and, after that period, shall pay such amount of license fee, for further licenses as the Corporation, with permission of the Legislature, shall then prescribe; or, in default of consenting thereto, shall surrender the road, with all the equipments and appurtenances thereto belonging, to the said Corporation, at a fair and just valuation of the same.

THIRTEENTH. Within a reasonable time after the passing of this resolution, the said associates, or a majority in interest thereof, shall form themselves into a joint-stock association, which association shall be vested with all the rights and privileges hereby granted, and shall have power, by the votes of at least a majority in interest of the associates, to frame and establish articles of association and by-laws providing for the construction, operation, and management of the said railway, the mode of admitting new associates, and of transferring the shares or interests of any of the associates to new associates or assigns, the number, duties, mode of appointment, tenure, and compensation of officers, the manner of making contracts, amending the by-laws, and calling in assessments from the associates, and generally the means and mode of establishing the railway and carrying it on, and of controlling and managing the property and affairs of the said association.

FOURTEENTH. The association shall not be deemed dissolved by the death or act of any associate, but his successor in interest shall stand in his place, and the rights of each associate shall depend on his own fulfillment of the conditions imposed on him by these restrictions, or the articles of association and by-laws of the association ; and, in case of his failure to fulfill the same, after 20 days' notice in writing to him so to do, his rights shall be forfeited to and devolve upon the remaining associates ; and said associates may at any time incorporate themselves under the General Railroad Act, whenever two-thirds in interest of the associates shall require it.

FIFTEENTH. The associates, whose names are set forth in this resolution, shall, by writing, filed with the clerk of the Common Council, signify their acceptance thereof, and agree to conform thereto ; and all new associates or assigns duly admitted, according to the provisions of the articles of association and by-laws, shall be deemed parties to such agreement.

Adopted by the Board of Aldermen, Nov. 19, 1852.

Adopted by the Board of Assistants, Dec. 6, 1852.

Board of Aldermen, Dec. 18, 1852, received from his Honor the Mayor, with his objections thereto.

Board of Aldermen, Dec. 29, 1852, taken up, reconsidered and adopted, notwithstanding the objections of his Honor the Mayor thereto, a majority of all the members elected voting therefor.

Board of Assistants, Dec. 30, 1852, taken up, reconsidered, and adopted, notwithstanding the objections of the Mayor thereto, a majority of all the members elected voting therefor ; therefore, under the provisions of the amended charter, the same became adopted.

SIXTH AVENUE RAILROAD.

AGREEMENT MADE, this sixth day of September, in the year one thousand eight hundred and fifty-one, between the Mayor, Aldermen and Commonalty of the city of New York, parties of the first part, and the persons named in the resolutions hereinafter set forth, who shall duly sign and execute this agreement, and their successors, associates and assigns, duly becoming parties thereto, as hereinafter provided, of the second part.

Whereas, The said parties of the first part, in Common Council convened, did, on the fourth day of June, one thousand eight hundred and fifty-one, duly pass and adopt the following resolutions, which were afterwards, and on the thirtieth day of July, in said year, duly signed by the Mayor of said city, and became operative and binding, in the words and figures following :

Resolved, That the persons to whom permission is granted by the following resolutions, and those who may hereafter become associated with them, have the authority and consent of the Common Council to lay a double track for a railroad in the following streets, viz. : from a point at the intersection of Chambers street and West Broadway, thence along West Broadway to Canal street, thence along and down Canal street to Hudson street, along Hudson street and 8th avenue to a point at or near 51st street ; and that said railroad be continued through the 8th avenue to Harlem river, whenever required by the Common Council, and as soon and as fast as said avenue is graded, upon the following stipulations and conditions, viz. : Such track or tracks to be laid under the direction of the Street Commissioner, and on such grades as are now established, or may hereafter be established, by the Common Council, the said parties to become bound, in a sufficient penalty, to keep in good repair the space between the track and the space outside the same, on either side, of at least eight feet in width, of each · street in which the rails are laid, and also that no motive power, excepting horses, be used below 51st street, and upon the further condition that said parties shall place new cars on said railroad, with all the modern improvements, for the convenience and comfort of passengers, and that they run cars thereon, each and every day, both ways, as often as the public convenience may require, under such directions as the Street Commissioner and Common Council may from time to time prescribe.

And provided, also, that the said parties shall, in all respects, comply with the direction of the Street Commissioner and of the Common Council in the building of said railroad, and in the running of the cars thereon, and in any other matter connected with the regulation of said railroad.

And provided, also, that the said parties shall, before this permission takes effect, enter into a good and sufficient agreement with the Mayor, Aldermen, and Commonalty of the city of New York, to be drawn and approved of by the Counsel to the Corporation, themselves to abide by and perform the stipulations and provisions herein contained, and also all such other resolutions, or ordinances, as may be passed by the Common Council, relating to the said road.

And further, that they run a car thereon each and every day, both ways, as often as every 15 minutes, from 5 to

6 o'clock, A. M., every 4 minutes, from 6 A. M. to 8 P. M.,
every 15 minutes, from 8 P. M. to 12 P. M., and every 30
minutes, from 12 P. M. to 5 o'clock A. M., and as much
oftener as public convenience may require, under such
directions as the Common Council may from time to time
prescribe.

Also, that the rate of passage on said railroad shall
not exceed a greater sum than 5 cents for the entire
length of said road ; and also that the Common Council shall have the power to cause the same, or any
part thereof, to be taken up at any time they may
see fit ; and also, that the said parties, or either of them,
shall not assign their interest in the said road without first
obtaining the consent of the Common Council thereto :
also, that such track or tracks shall be laid upon a foundation of concrete, with a grooved rail, or such other rail as
may be approved of by the Street Commissioner, even with
the surface of the streets through which they may pass, and
shall be commenced within three months, and completed
to Fifty-first street within one year, and from Fifty-first
street to the Harlem River within three years, from the
passage of this resolution ; also, that the foundation on
each side of the rails shall be paved with square grooved
blocks of stone, similar to the Russ pavement, as far up as
Fifty-first street ; that the said parties are to keep an account of the receipts of each road monthly, and report the
same to the Comptroller, monthly, under oath ; that the
said parties shall connect their road with such other roads
as the Common Council may order to be connected therewith ; that they shall file with the Comptroller a statement,
under oath, of the cost of each mile of road completed,

and agree to surrender, convey, and transfer the said road
to the Corporation of the city of New York, whenever required so to do, on payment, by the Corporation, of the
cost of said road, as appears by said statement, with ten
per cent. advance thereon.

That said parties, on being required at any time by the
Corporation, and to such extent as the Common Council
shall determine, shall take up, at their own expense, said
rails, or such part thereof as they shall be required, and on
failure so to do, in ten days after such requirement, the
same may be done, at their expense, by the Street Commissioner.

Resolved, That the persons, to whom permission is grant-
by the following resolutions, have the authority and consent of the Common Council to lay a single track in the
following streets: Commencing at the corner of Chambers
street and West Broadway, through Chambers street to
Church street, through Church street to Canal street,
through Canal street to Wooster street, through Wooster
to Fourth street, to Thompson street, with a single track;
thence, with a double track, through Fourth street and
Sixth avenue, to Harlem; also to lay a single track in
Thompson street, from Fourth to Canal street, to connect with the Eighth Avenue Railroad, and extend the
same up the Sixth avenue to Harlem River, whenever
required by the Common Council, and as soon and as
fast as said avenue is graded sufficiently to permit such
track to be laid, upon the same terms, stipulations,
and conditions, as are provided in the annexed resolutions
in relation to the railroad on the Eighth avenue, except
that no motive power except horses shall be used below

Forty-second street; that said railroad upon the Sixth avenue shall be commenced within three months, and completed to Forty-second street within one year, and from Forty-second street to the Harlem River within three years, from the passage of this resolution; also, that the foundation on each side of the rails shall be paved with square grooved blocks of stone, similar to the Russ pavement, as far up as Thirty-second street, and that such parts of the Eighth Avenue road as may be used by the Sixth Avenue road, from the connection in Canal street and West Broadway to Chambers street, shall be built at the joint expense of said Sixth and Eighth Avenue Railroads.

Resolved, That each of said passenger-cars, to be used en said roads, shall be annually licensed by the Mayor; and there shall be paid annually for such license such sum as the Common Council may hereafter determine.

Resolved, That the permission granted to lay or build a railroad in the following streets, viz.: commencing at a point at the intersection of West Broadway and Chambers street, thence through Chambers street to Church street, through Church street to Canal street, and through Canal street to Wooster street, through Wooster street to Fourth street, with a single track; thence through Fourth street to Sixth avenue, and through Sixth avenue to Harlem, with a double track, also, to lay a single track in Thompson street, from Fourth street to Canal street, to connect with the Eighth Avenue Railroad, be given to James S. Libby, George R. Howell, William Flagg, William H. Adams, John Post, jun., Edmund Morris, Matthew D. Greene, John Ridley, William Ebbitt, Ward, Bolster & Jacacks and Finch, Sanderson and Beers.

And whereas, said parties of the first part, on the said fourth day of June, one thousand eight hundred and fifty-one, in Common Council convened, did duly pass and adopt certain other resolutions, which were likewise duly signed and approved by the said Mayor on the said thirtieth day July, one thousand eight hundred and fifty-one, and became operative and binding, providing for the laying or building of another railroad, designated as the Eighth Avenue Railroad, provided for in said resolutions hereinbefore set forth ; and further providing and directing that such parts of the Eighth Avenue Railroad as may be used by the Sixth avenue road, from the connection in Canal street and West Broadway to Chambers street, should be built at the joint expense of the Sixth and Eighth avenue roads.

And whereas, it is deemed necessary by the said parties of the first part, in order to preserve and duly effectuate the grants, objects, stipulations and intentions of the said resolutions, and for the purpose of more specifically determining the interest of said parties in the rights and privileges granted by said resolutions, that provision should be made for an organization or association between said parties of the second part, their successors, associates and assigns, duly admitted according to this agreement, defining the mode in which the necessary capital for building the said railroad shall be contributed, and the manner in which the construction and management of the said railroad shall be conducted and controlled.

Now, it is hereby mutually declared that the separate and individual interest of any or either of the said parties of the second part, their successors, associates and assigns, in the said grant, and all licenses, rights, privileges and powers conferred or provided for in the said resolutions, shall be conditioned and dependent upon the strict observance, performance, and fulfilment, by such person, of the terms of said resolutions, and of this agreement; and that, in case of failure to perform the same and every part thereof, said grant shall be inoperative as to such person so failing, and his interest therein shall cease and determine— said grant remaining operative, in every respect, as to all other of said parties, their successors, associates and assigns. And it is hereby covenanted, agreed and declared, by and between the parties aforesaid, as follows, viz.:

First. The said parties of the second part, for themselves and their successors, associates and assigns, do hereby covenant and agree with the said parties of the first part, and with each other, that they will well and truly observe, perform, fulfill and keep the said resolutions hereinbefore particularly set forth, and all and every the provisions, stipulations, restrictions and conditions therein contained and thereby imposed, according to the true intent and meaning thereof, it being understood that the rate of passage on said road shall not exceed five cents for any distance, and also that the road shall be completed at the times and in the manner stated in said resolutions.

Second. The said parties of the second part, to the end that the provisions and intentions of the said resolutions may be fully carried into effect, the interests of the respective parties definitely ascertained, and the manner in which the construction and management of said road shall be conducted and controlled effectually defined, do further

18

covenant and agree with the said parties of the first part, and with each other, to associate and organize themselves together, in the manner and upon the terms and conditions following, viz. :

Within ten days after this agreement is duly executed, the said parties of the second part, unless they, or a majority of them, shall have previously organized themselves to the same effect as herein provided, shall and will organize themselves into an association or company, to be called The Sixth Avenue Railroad Company, for the purpose of constructing, operating and managing said railroad ; the first meeting of the said parties to be called by the clerk of the Common Council, who shall, within three days after the due execution of this agreement, give, or cause to be given, a notice in writing, delivered to the persons composing the said parties of the second part personally, or left at their residences or places of business, specifying the time and place when and where such meeting shall be held. The said parties of the second part, or as many of them as shall meet in pursuance of said notice, shall thereupon proceed, as before provided, to organize themselves into the said company, and shall have power and authority, by the votes of a majority of the parties so assembled,

1. To estimate and declare the amount of capital requisite to construct the said railroad, provide cars, motive power, stations, buildings, fixtures, and for all other expenses requisite to put the said railroad into thorough practical operation.

2. To prescribe the mode in which said capital, and all other sums that may thereafter be required for the business of said company, shall be subscribed for, and the time or times when the same shall be paid, and the manner in which shares and interests of the parties refusing or neglecting to subscribe or to pay may be forfeited.

3. To adopt suitable resolutions, by-laws, rules and regulations for the organization of said company, the subscription and payment of its capital and all other sums that may thereafter be required for its construction, operation and future business, the execution of contracts, the liability of members, the term, compensation, accountability, election, removal and duties of its officers, the disbursement of moneys, the transfer or assignment of the shares of its members, and the entire management, direction and control of its affairs, business, property and officers. Such by-laws may be altered from time to time, in the manner prescribed therein.

THIRD. The said parties of the second part shall be entitled to subscribe equally for the amount proposed as the original capital stock of said company, and if any of them neglect to subscribe, or shall subscribe less than his proportion, the others may subscribe equally for the remainder, so as to make up a subscription for the whole amount. If, for any reason, it shall be requisite to make other subscriptions, the persons who shall then be members of said company shall be entitled to subscribe for the amount so required, in proportion to the amounts of capital stock then held by them, and if any shall neglect to subscribe, or shall subscribe less than his proportion, the others may subscribe equally for the remainder.

FOURTH. Every person refusing or neglecting to subscribe to the capital stock of said company as originally declared, or to any subsequent increase thereof, or to pay his subscription or any instalment thereof, at the times prescribed at the first meeting of said company as aforesaid, or by the resolutions or by-laws of the said company, all his rights, powers and privileges under said grant of the parties of the first part, and all his interest therein, shall be deemed to be freely and voluntarily waived and abandoned for the benefit of said company and its remaining members, and shall cease, determine and be utterly null and void, and he shall no longer be a member of said company, nor have any voice in the management of its afiairs, nor have any title or interest in its property. But such waiver or abandonment shall not be deemed to have taken place until twenty days shall have elapsed after such person shall have had written notice of the required subscription or payment. But such person may, by a resolution duly adopted by the said company, be reinstated in any or all of the rights, privileges and advantages so as aforesaid waived and lost, but upon such terms and conditions as may be thereby provided.

FIFTH. Every person who shall become a member of said company shall thereby become a party to this agreement and all its conditions and stipulations, and the company may direct the mode by which future members shall become so obligated, and no person shall become a member except on condition of becoming so obligated by agreement in writing duly executed.

SIXTH. The said railroad grant, property, rights and appurtenances shall belong to and be the property of the persons who, for the time being, shall compose the said Sixth Avenue Railroad Company, in proportions equivalent to their shares of said capital stock; subject, however, to the management of the same in the manner herein provided.

SEVENTH. Any shareholder may transfer his shares or interest, after he shall have paid one-third of his original subscription, on procuring the consent of a majority in interest of the shareholders, expressed by resolution duly adopted; subject, however, to the provisions of this agreement, and on such terms and conditions as the by-laws may prescribe.

EIGHTH. This Company shall not be dissolved by the death or insolvency of any of its members, nor by act or operation of law, but in such and the like cases shall continue, and the persons becoming lawfully entitled to the shares shall become members of the said Company, and said Company shall have authority to incorporate themselves under the general railroad act, whenever two-thirds in interest of the shareholders shall require the same.

In witness whereof, to one part of these presents, remaining with the said parties of the first part, the said parties of the second part have affixed their hands and seals; and to the other part thereof, remaining with the said parties of the second part, the said parties of the first part have caused the common seal of the city of New York to be affixed, the day and year first above written.

JAMES S. LIBBY,   [L. S.]
GEORGE R. HOWELL,   [L. S.]
WILLIAM FLAGG,   [L. S.]
WM. H. ADAMS,   [L. S.]
JOHN POST, Jr.,   [L. S.]
EDMUND MORRIS,   [L. S.]
MATTHEW DAVIS GREENE,   [L. S.]
JOHN RIDLEY,   [L. S.]
WILLIAM EBBETT,   [L. S.]

Sealed and delivered in presence of
    H. H. ANDERSON,
as to Libby, Howell, Flagg, Greene, Ebbett, Adams, and Morris;
    In presence of CARLTON EDWARDS, as to John Post, Jr.; and
    BERNARD J. MALONE, as to J. Ridley.

CITY AND COUNTY OF NEW YORK, ss. :—On the ninth day of September, one thousand eight hundred and fifty-one, before me personally came James S. Libby, George R. Howell, William Flagg, William H. Adams, Matthew D. Greene, and William Ebbett; and on the fifteenth day of September, in the same year, before me personally came Edmund Morris, all of whom were known to me to be persons described in and who executed the foregoing instrument, and severally acknowledged that they executed the same; and on the nineteenth day of September, one thousand eight hundred and fifty-one, before me personally came Carlton Edwards, a subscribing witness to said instrument, to me known, who, being by me duly sworn, did depose and say that he resides in the city of New York, that he knew John Post, jr., and knew him to be one of the persons described in and who executed the foregoing instrument; that he saw him sign the same, and that he acknowledged in his presence that he executed the same, and subscribed his name as a witness thereto. And on the same day before me personally came Bernard J. Malone, one of the subscribing witnesses to said instrument, to me known, who, being by me duly sworn, did depose and say that he resides in the city of New York, that he knew John Ridley, and knew him to be one of the persons described in and who executed the foregoing instrument, that he saw him sign the same, that he acknowledged in his presence that he executed the same, and that he subscribed his name as a witness thereto.

HENRY H. ANDERSON,

Commissioner of Deeds.

Resolved, That the route of the Sixth Avenue Railroad be, and the same is hereby, changed from the present location so as to run as follows, viz.: Commencing at the intersection of Chambers street and West Broadway, running thence, with a double track, through West Broadway to Canal street, through Canal street to Varick, through Varick to Carmine street, through Carmine street and the Sixth avenue, to intersect with the original grant of the Sixth Avenue Railroad Company.

Resolved, That the portion of said railroad track to be laid down in West Broadway, from Chambers street to Canal street, shall be built jointly by the Sixth and Eighth Avenue Railroad Companies, and in the event of either party refusing or neglecting to unite in the construction of such portion of the road, or pay for their proportionate share of the expense thereof, then it shall be competent for either of said parties to proceed with the work at their own expense and for their exclusive benefit, until the other party shall actually pay their proportion of such expense ; further, should any difficulty arise between said Sixth and Eighth Avenue Railroad Companies as to the cost and value of building said road and rights to run over the same, each party shall have the privilege to select a referee, not interested in any wise in either of said roads, to adjust all difficulties, but should said referees not be able to make a proper and amicable settlement of any dispute arising between the parties hereinbefore mentioned, then it shall be competent for the Common Council to select a third person as referee, who shall investigate the subject-matter in dispute, and a decision from a majority of said referees shall be final and conclusive ; nothing, however, contained in this resolution shall be construed to interfere with the power and authority of the Common Council to prescribe rules and regulations, from time to time, for the control and management of said railroad.

Adopted by the Board of Assistants, June 9, 1852.

Adopted by the Board of Aldermen, June 17, 1852.

Approved by the Mayor, June 23, 1852.

Petition of James S. Libby, President of the Sixth

Avenue Railroad Company, to grant the Manhattan Gas-Light Company permission to lay their mains in the Sixth avenue to Forty-fourth street, so that the depot may be lighted with gas, was granted

By the Board of Assistants, Nov. 9, 1852.

By the Board of Aldermen, Nov. 17, 1852.

Approved by the Mayor, Nov. 19, 1852.

Resolved, That the Eighth Avenue Railroad Company have authority, and the privilege is hereby granted to them, to extend their rail (which is to be constructed in like manner as their present) through Canal street to Broadway, and also from its present termination at Chambers street, t h College place to Barclay street, and through Barclay and Church streets, or across Barclay street and through the buildings which they have rented or procured, or may rent or procure, for the purpose, to and into Vesey street ; through Vesey street to Broadway, and through Church street, from Vesey street to Chambers street, and through Chambers street to its present termination aforesaid, and to run their cars over the same ; and when they shall have made such extension, then and thereafter they shall be at liberty to charge every passenger, who may come to and ride any distance upon any part of their road below Fifty-first street, five cents for riding on that part of their road. Anything in the resolutions or agreement, under which the said company are now acting, inconsistent with any of the privileges granted by this resolution, is hereby modified so as to conform thereto.

And also, *Resolved*, That the Sixth Avenue Railroad Company, upon paying to the said Eighth Avenue Railroad Company the half part of the cost of that portion of their road lying between Varick street and West Broadway, and of keeping it in repair from time to time hereafter, and also the half of the costs and of the repairs, from time to time, of the extensions above authorized, shall be at liberty to use and own half of the same, and run their cars thereon, and to charge every passenger, who may come to ride any distance upon any part of their road below Forty-third street, the like sum of five cents for riding on that part of their road ; and anything in the resolutions or agreement, under which they are now acting, inconsistent with the said privileges, is hereby modified so as to conform to this resolution.

Adopted by the Board of Assistants, Nov. 11, 1852.

Adopted by the Board of Aldermen, Nov. 22, 1852.

Received from his Honor the Mayor, Dec. 13, 1852, without his approval or objections thereto ; therefore, under the provisions of the amended charter, the same became adopted.

Whereas, A resolution has passed the Common Council, and which resolution was permitted to become a law by the non-action of the Mayor, authorizing the Sixth and Eighth Avenue Railroad Companies to extend their road through College place to and across Barclay street, or through Barclay or Vesey street to Broadway, and to return through Chambers street to its present termination ; and

Whereas, It was the understanding, although not specified in the resolution, that said extension should and was to be only with a single track, yet the said Eighth Avenue Railroad Company is now engaged in constructing said road with a double track, which will have the effect to shut from said street all other travel, and be a serious detriment to the interests of the citizens generally, and to the owners of property on the line especially ; and, as there seems to be a very serious misunderstanding as to whether there should be constructed a single or a double track, therefore, be it

Resolved, That the Street Commissioner be, and he is hereby, directed to cause all further proceedings of said Companies, in the construction of said roads through the streets mentioned in the preamble, to be stayed and suspended until the further action of the Common Council, except they build a single track only through said streets, and if that course be not adopted by said Companies, that then the said Street Commissioner be, and he is hereby, directed to have said streets restored to their former condition without delay.

Adopted by the Board of Aldermen, Dec. 18, 1852.

Adopted by the Board of Assistants, Dec. 20, 1852.

Approved by the Mayor, Dec. 21, 1852.

Resolved, That the Commissioner of Streets be, and he is hereby, directed to notify the Sixth Avenue Railroad Company to have the railroad alley, between Vesey and Barclay streets, closed up with suitable gates forthwith.

Adopted by the Board of Councilmen, April 24, 1854.

Adopted by the Board of Aldermen, May 19, 1854.

Approved by the Mayor, May 26, 1854.

Resolved, That the Sixth Avenue Railroad Company be, and they are hereby, directed to complete their track to Fifty-fourth street without delay, in accordance with the ordinance heretofore adopted by the Common Council, on or before the 1st of June, 1857, otherwise the Counsel to the Corporation is authorized and directed to commence proceedings against them, to compel them to complete said track.

Adopted by the Board of Aldermen, March 9, 1857.

Adopted by the Board of Councilmen, April 17, 1857.

Approved by the Mayor, April 20, 1857.

The President of the Croton Aqueduct Board directed to notify the Sixth Avenue Railroad Company to put in good repair, forthwith, all the pavements in and about their rails, and, in case of neglect or refusal, to have the same done at the expense of the Sixth Avenue Railroad Company.

Adopted by the Board of Aldermen, Sept. 10, 1857.

Adopted by the Board of Councilmen, Sept. 14, 1857.

Approved by the Mayor, Sept. 16, 1857.

Resolved, That the Sixth Avenue Railroad Company be, and they are hereby, directed to repair their track and keep the same in good order on the established grades of the streets and avenues in which their said track is laid, in accordance with the provisions of the charter granted them by the Common Council, and that the same be done within fifteen days after the approval of this resolution by his Honor the Mayor, and in case said Sixth Avenue Railroad Co. fail to comply with the provisions of this resolution, then, that the Croton Aqueduct Board be, and they are hereby, authorized and directed to cause the track of the said Sixth Avenue Railroad Co. to be repaired and put in good order, and charge the expense thereof to the Sixth Avenue Railroad Company.

Adopted by the Board of Aldermen, May 11, 1859.

Adopted by the Board of Councilmen, June 27, 1859.

Approved by the Mayor, July 28, 1859.

Resolved, That the contractor for paving Sixth avenue, between Carmine and Forty-second streets, with square-block pavement, be, and he is hereby, directed not to pave that portion of the avenue between the outside rails of the Sixth Avenue Railroad, but to continue only the paving of the avenue between the outside rails and the curbs on either side of said Sixth avenue in accordance with the contract.

Resolved, That the Sixth Avenue Railroad Company be, and they are hereby, authorized and permitted, at their own expense, to pave the space between and within the outside rails of their said road with small cobble-stones, and are hereby directed to keep the same in good order and condition, all of which to be done under the direction of the Croton Aqueduct Department.

Resolved, That the Croton Aqueduct Department be, and are hereby, requested and directed to cause the aforesaid resolutions to be carried into effect.

Adopted by the Board of Councilmen, Oct, 20, 1859.

Adopted by the Board of Aldermen, Oct. 20, 1859.

Approved by the Mayor, Oct. 27, 1859.

SEVENTH AVENUE RAILROAD.

Whereas, A bill has been introduced in the Assembly of this State, to authorize John A. Kennedy and others to construct a railroad, with a double track, upon and along the Seventh avenue at Fifty-ninth street; thence along Seventh avenue to Broadway; thence along Broadway and Union place to University place; along University place to Sixth street; thence along Sixth street to Greene street; thence along Greene street to Canal street; thence along Canal street to West Broadway; thence along West Broadway and College place to Park place; thence along Park place to Church street; thence along Church street to Fulton street; thence along Fulton street to Broadway; also along Park place to Broadway, and thence back through Park place to Church street, and thence through Church street to Canal street; thence along Canal street to Mercer street; thence along Mercer street to Waverley place (or Sixth street), to connect with the track in University place; and

Whereas, It is believed that no public emergency requires any such railroad, but, on the contrary, the citizens of New York, and particularly those living on the route aforesaid, are wholly opposed to the measure; therefore be it

Resolved, That this Common Council do most earnestly remonstrate against the said project, and is opposed to the passage of any law which tends to obstruct the streets, avenues or highways of the city by laying rail-tracks therein.

Adopted by the Board of Councilmen, April 9, 1857.

Adopted by the Board of Aldermen, April 10, 1857.

Approved by the Mayor, April 11, 1857.

EIGHTH AVENUE RAILROAD.

Resolved, That the annexed form of agreement is approved as the form of agreement to be executed by the parties in whose favor resolutions were heretofore adopted by the Common Council, granting permission to lay or build the railroad known as the Eighth Avenue Railroad, and that the said parties be, and they are hereby, required to duly sign and execute said agreement within ten days after the same shall have been prepared for execution by the Counsel to the Corporation ; and that any or either of the said parties, who shall neglect or refuse to sign and execute said agreement within the time aforesaid, shall be deemed to have waived and forfeited all and every right, benefit and advantage under the resolutions heretofore passed, granting said permission to and for the benefit of the persons named in the same, who duly execute said agreement in manner aforesaid, and the said permission shall take effect accordingly.

Resolved, That an agreement, in the like form as that

annexed, varied and adapted to the resolutions heretofore
passed, granting permission to construct the Sixth Avenue
Railroad, be prepared by the Counsel to the Corporation,
and that the parties to whom such permission is granted
be, and they are hereby, required to sign and execute said
agreement within ten days after the same shall have been
prepared for execution, as aforesaid, and that any or either
of the said parties neglecting or refusing to execute said
agreement, in manner aforesaid, shall be deemed to have
waived and forfeited all and every right, benefit and advantage under said resolutions, to and for the benefit of those
named in the said resolutions, who duly execute said agreement, and said permission shall take effect accordingly.

Adopted by the Board of Aldermen, Sept. 3, 1851.

Adopted by the Board of Assistants, Sept. 4, 1851.

Approved by the Mayor, Sept. 5, 1851.

AGREEMENT, made this sixth day of September, one
thousand eight hundred and fifty-one, between the Mayor,
Aldermen and Commonalty of the city of New York, parties
of the first part, and the persons named in the resolutions
hereinafter set forth, who shall duly sign and execute this
agreement, and their successors, associates and assigns, duly
becoming parties thereto, as hereinafter provided, of the
second part:

Whereas, The said parties of the first part, in Common
Council convened, did, on the fourth day of June, one
thousand eight hundred and fifty-one, duly pass and adopt
the following resolutions, which were afterwards and on

the thirtieth day of July in said year duly signed and approved by the Mayor of said city, and became operative and binding in the words and figures following :

Resolved, That the persons to whom permission is granted by the following resolutions, and those who may hereafter become associated with them, have the authority and consent of the Common Council to lay a double track for a railroad in the following streets, viz. : from a point at the intersection of Chambers street and West Broadway, thence along West Broadway to Canal street, thence along and down Canal street to Hudson street, along Hudson street to Eighth avenue, to a point at or near 51st street and that said railroad be continued through the Eighth avenue to Harlem River, whenever required by the Common Council and as soon and as fast as said avenue is graded, upon the following stipulations and conditions, viz.:

Such track or tracks to be laid under the direction of the Street Commissioner and on such grades as are now established or may hereafter be established by the Common Council, the said parties to become bound in a sufficient penalty to keep in good repair the space between the track and the space outside the same on either side, of at least eight feet in width of each street in which the rails are laid, also that no motive power, excepting horses, be used below 51st street, and upon the further condition that said parties shall place new cars on said railroad, with all the modern improvements for the convenience and comfort of passengers, and that they run cars thereon each and every day, both ways, as often as the public convenience may re-

19

quire, under such directions as the Street Commissioner and Common Council may, from time to time, prescribe.

3. *And provided also*, that the said parties shall, in all respects, comply with the direction of the Street Commissioner and of the Common Council in the building of said railroad and in the running of the cars thereon, and in any other matter connected with the regulation of said railroad.

4. *And provided also*, that the said parties shall, before this permission takes effect, enter into a good and sufficient agreement with the Mayor, Aldermen and Commonalty of the city of New York, to be drawn and approved of by the Counsel to the Corporation, binding themselves to abide by and perform the stipulations and provisions herein contained and also all such other resolutions or ordinances as may be passed by the Common Council relating to the said road.

5. *And further*, that they run a car thereon each and every day, both ways, as often as every fifteen minutes from 5 to 6 o'clock A. M., and every four minutes from 6 o'clock A. M., to 8 o'clock P. M., every fifteen minutes from 8 o'clock P. M. to 12 o'clock P. M., and every thirty minutes from 12 o'clock P. M., to 5 o'clock A. M.; and as much oftener as public convenience may require, under such directions as the Common Council may, from time to time, prescribe.

6. *Also*, that the rate of passage on said railroad shall not exceed a greater sum than five cents for the entire length of said road, and also that the Common Council shall have

the power to cause the same, or any part thereof, to be taken up at any time they may see fit; and also that the said parties, or either of them, shall not assign their interest in the said road without first obtaining the consent of the Common Council thereto; also that such track or tracks shall be laid upon a foundation of concrete, with a grooved rail or such other rail as may be approved of by the Street Commissioner, even with the surface of the streets through which they may pass, and shall be commenced within three months and completed to 51st street within one year, and from 51st street to Harlem River within three years, from the passage of this resolution; also that the foundation on each side of the rails shall. be paved with square grooved blocks of stone, similiar to the Russ pavement, as far up as 51st street; that the said parties are to keep an account of the receipts of each road monthly and report the same to the Comptroller monthly, under oath; that the said parties shall connect their road with such other roads as the Common Council may order to be connected therewith; that they shall file with the Comptroller a statement, under oath, of the cost of each mile of road completed, and agree to surrender, convey and transfer the said road to the Corporation of the city of New York, whenever required so to do, on payment, by the Corporation, of the cost of said road, as appears by said statement, with ten per cent. advance thereon; that said parties, on being required at any time by the Corporation, and to such extent as the Common Council shall determine, shall take up, at their own expense, said rails, or such part thereof as they shall be required, and on failure so to do, in ten days after such requirement, the same may be done at their expense by the Street Commissioner.

Resolved, That each of said passenger cars, to be used on said road, shall be annually licensed by the Mayor; and there shall be paid, annually, for such license, such sum as the Common Council may hereafter determine.

Resolved, That the permission granted to lay or build a railroad track in the following streets, viz: from a point at the intersection of West Broadway and Chambers street, thence through West Broadway to Canal street, down Canal street to Hudson street, along Hudson street and Eighth avenue to Harlem River, be granted and given to John Pettigrew, Edmund R. Sherman, Solomon Kipp, Abraham Brown, Washington Smith, Joseph N. Barnes, John O'Keefe, John J. Duryea, Jesse A. Marshall and Timothy Townsend.

And whereas, Said parties of the first part, on the said fourth day of June, one thousand eight hundred and fifty-one, in Common Council convened, did duly pass and adopt certain other resolutions, which were likewise duly signed and approved by the said Mayor, on the said thirtieth day of July, one thousand eight hundred and fifty-one, and became operative and binding, providing for the laying or building of another railroad, designated as the Sixth Avenue Railroad, provided for in said resolutions hereinbefore set forth; and further providing and directing that such parts of the Eighth Avenue Railroad as may be used by the Sixth Avenue Road, from the connection in Canal street and West Broadway to Chambers street, should be built at the joint expense of the said Sixth and Eighth Avenue Roads.

And whereas, It is deemed necessary by the said parties

of the first part, in order to preserve and duly effectuate the grants, objects and stipulations and intentions of the said resolutions, for the purpose of more specifically determining the interests of the said parties in the rights and privileges granted by said resolutions, that provision should be made for an organization or association between the said parties of the second part, their successors, associates and assigns, duly admitted, according to this agreement, defining the mode in which the necessary capital for building the said railroad shall be contributed, and the manner in which the construction and management of the said railroad shall be constructed and controlled. Now, it is hereby mutually declared that the separate or individual interest of any or either of the said parties of the second part, their successors, associates and assigns, in the said grant, and all licenses, rights, privileges and powers, conferred or provided for in the said resolutions, shall be conditioned and dependent upon the strict observance, performance and fulfilment by such person of the terms of said resolutions and of this agreement; and that, in case of failure to perform the same, and every part thereof, said grant shall be inoperative as to such person so failing, and his interest therein shall cease and determine; said grant remaining operative in every respect as to all others of said parties, their successors, associates and assigns. And it is hereby covenanted, agreed and declared, by and between the parties aforesaid, as follows, viz.:

First. The said parties of the second part, for themselves and their successors, associates and assigns, do hereby covenant and agree with said parties of the first part, and with each other, that they will well and truly observe, perform, fulfill and keep the said resolutions hereinbefore particularly set forth, and all and every the provisions, stipulations, restrictions and conditions therein contained and thereby imposed, according to the true intent and meaning thereof, it being understood that the rate of passage on said road shall not exceed five cents for any distance; and, also, that the said road shall be completed at the times and in the manner stated in said resolutions.

SECOND. The said parties of the second part, to the end that the provisions and intentions of the said resolutions may be fully carried into effect, the interest of the respective parties definitely ascertained, and the manner in which the construction and management of said road shall be conducted and controlled effectually defined, do further covenant and agree with the said parties of the first part, and with each other, to associate and organize themselves in the manner and upon the terms and conditions following, viz. :

Within ten days after this agreement is duly executed the said parties of the second part, unless they, or a majority of them, shall have previously organized themselves to the same effect, as herein provided, shall and will organize themselves into an association or company, to be called the Eighth Avenue Railroad Company, for the purpose of constructing, operating and managing said railroad, the first meeting of said parties to be called by the Clerk of the Common Council, who shall, within three days after the due execution of this agreement, give, or cause to be given, a notice in writing, delivered to the persons composing the said parties of the second part, personally or left at their residences or places of business, specifying the time and place, when and where such meeting shall be held.

The said parties of the second part, or as many of them as shall meet in pursuance of said notice, shall thereupon proceed, as before provided, to organize themselves into the said company, and shall have power and authority, by the votes of a majority of the parties so assembled,

1. To estimate and declare the amount of capital requisite to construct the said railroad, provide cars, motive power, stations, buildings, fixtures, and for all other expenses requisite to put the said railroad into thorough practical operation.

2. To prescribe the mode in which said capital and all other sums that may thereafter be required for the business of said company, shall be subscribed for, and the time or times when the same shall be paid in, and the manner in which the shares and interests of the parties refusing or neglecting to subscribe or to pay, may be forfeited.

3. To adopt suitable resolutions, by-laws, rules and regulations for the organization of said company, the subscription and payment of its capital and all other sums that may thereafter be required for its construction, operation and future business, the execution of contracts, the liability of members, the term, compensation, accountability, election, removal and duties of its officers, the disbursement of moneys, the transfer or assignment of shares of its members, and the entire management, direction and control of its affairs, business, property and officers. Such by-laws may be altered, from time to time, in the manner prescribed therein.

THIRD. The said parties of the second part shall be entitled to subscribe equally for the amount proposed as the original capital stock of said company; and if any of them neglect to subscribe, or shall subscribe less than his proportion, the others may subscribe equally for the remainder, so as to make up a subscription for the whole amount. If, for any reason, it shall be requisite to make other subscriptions, the persons who shall then be members of said company shall be entitled to subscribe for the amount so required in proportion to the amount of capital stock then held by them, and if any shall neglect to subscribe, or shall subscribe for less than his proportion, the others may subscribe equally for the remainder.

FOURTH. Every person refusing or neglecting to subscribe to the capital stock of said company, as originally declared, or to any subsequent increase thereof, or to pay his subscription or any installment thereof, at the times prescribed at the first meeting of said company, as aforesaid, or by the resolutions or by-laws of the said company, all his rights, powers and privileges, under said grant of the parties of the first part, and all his interest therein, shall be deemed to be freely and voluntarily waived and abandoned, for the benefit of said company and its remaining members, and shall cease, determine and be utterly null and void, and he shall no longer be a member of said company, nor have any voice in the management of its affairs, nor any title or interest in its property; but such waiver and abandonment shall not be deemed to have taken place until twenty days shall have elapsed after such person shall have had written notice of the required subscription or payment. But such person may, by a resolution duly adopted by said company, be reinstated in any or all of the rights, privileges and advantages, &c., as aforesaid waived and lost, but upon such terms and conditions as may be thereby provided.

FIFTH. Every person who shall become a member of said company, shall thereby become a party to this agreement and all its conditions and stipulations, and the company may direct the mode by which future members shall become so obligated ; and no person shall become a member except on condition of becoming so obligated by agreement in writing, duly executed.

SIXTH. The said railroad grant, property, rights and appurtenances shall belong to and be the property of the persons who, for the time being, shall compose the said Eighth Avenue Railroad Company, in proportions equivalent to their shares of said capital stock, subject, however, to the management of the same, in the manner herein provided.

SEVENTH. Any shareholder may transfer his shares or interest after he shall have paid one-third of his original subscription, on procuring the consent of a majority in interest of the stockholders, expressed by a resolution duly adopted, subject, however, to the provisions of this agreement, and on such terms and conditions as the by-laws may prescribe.

EIGHTH. This company shall not be dissolved by the death or insolvency of any of its members, nor by act or operation of law, but in such and the like cases shall continue, and the persons becoming lawfully entitled to the shares shall become members of the said company ; and said company shall have authority to incorporate themselves under the general railroad act, whenever two-thirds in interest of the shareholders shall require the same.

In witness whereof, to one part of these presents, remaining with the said parties of the first part, the said parties of the second part have affixed their hands and seals; and to the other part thereof, remaining with the said parties of the second part, the said parties of the first part have caused the common seal of the city of New York to be affixed, the day and year first above written.

JOHN PETTIGREW,	[L. S.]
E. R. SHERMAN,	[L. S.]
JESSE A. MARSHALL,	[L. S.]
TIMOTHY TOWNSEND,	[L. S.]
JOHN O'KEEFE,	[L. S.]
JOHN J. DURYEA,	[L. S.]
WASHINGTON SMITH,	[L. S.]
SOLOMON KIPP,	[L. S.]
J. N. BARNES,	[L. S.]
ABRAHAM BROWN,	[L. S.]

Sealed and delivered in presence of HENRY H. ANDERSON, as to Sherman and Pettigrew, Marshall and Townsend and others.

CITY AND COUNTY OF NEW YORK, *ss.*—On the sixth day of September, one thousand eight hundred and fifty-one, before me personally came Edmund R.

Sherman ; on the eighth day of said month, before me personally came John Pettigrew ; on the tenth day of said month, before me personally came Jesse A. Marshall and Timothy Townsend ; on the thirteenth day of said month, before me personally came Washington Smith, Solomon Kipp, Joseph N. Barnes and Abraham Brown, all of whom are known to me to be persons described in and who executed the foregoing agreement ; and the said parties above named, severally upon the days above named, acknowledged that they executed said foregoing instrument.

HENRY H. ANDERSON,

Commissioner of Deeds.

———

Resolved, That the route of the Sixth Avenue Railroad be and the same is hereby changed from the present location so as to run as follows, viz. : commencing at the intersection of Chambers street and West Broadway, running thence with a double track through West Broadway to Canal street, through Canal street to Varick, through Varick to Carmine street, through Carmine street and the Sixth avenue to intersect with the original grant of the Sixth Avenue Railroad Company.

Resolved, That the portion of said railroad track to be laid down in West Broadway from Chambers street to Canal street shall be built jointly by the Sixth and Eighth Avenue Railroad Companies, and in the event of either party refusing or neglecting to unite in the construction of such portion of the road, or pay for their proportionate share of the expense therof, then it shall be competent for either of said parties to proceed with the work at their own expense and for their exclusive benefit until the other party shall actually pay their proportion of such expense; further, should any difficulty arise between said Sixth and Eighth Avenue Railroad Companies as to the cost and value of building said road and rights to run over the same, each party shall have the privilege to select a referee, not interested in any wise in either of said roads, to adjust all difficulties, but should said referees not be able to make a proper and amicable settlement of any dispute arising between the parties hereinbefore mentioned, then it shall be competent for the Common Council to select a third person as referee, who shall investigate the subject matter in dispute, and a decision from a majority of said referees shall be final and conclusive; nothing, however, contained in this resolution shall be construed to interfere with the power and authority of the Common Council to prescribe rules and regulations, from time to time, for the control and management of said railroad.

Adopted by the Board of Assistants, June 9, 1852.

Adopted by the Board of Aldermen, June 17, 1852.

Approved by the Mayor, June 23, 1852.

Resolved, That the Eighth Avenue Railroad Co. have authority, and the privilege is hereby granted to them, to extend their rail (which is to be constructed in like manner as their present) through Canal street to Broadway, and also from its present termination at Chambers street through College place to Barclay street, and through Barclay and Church streets, or across Barclay street through the buildings they have rented or procured, or may rent or procure, for the purpose, to and into Vesey street, through Vesey street to Broadway, and through Church street from Vesey street to Chambers street, and through Chambers street to its present termination aforesaid, and to run their cars over the same ; and when they shall have made such extension, then and thereafter they shall be at liberty to charge every passenger, who may come to and ride any distance upon any part of their road below 51st street, 5 cents for riding on that part of their road. Anything in the resolutions or agreement under which the said Company are now acting, inconsistent with any of the privileges granted by this resolution, is hereby modified, so as to conform thereto. And also,

Resolved, That the Sixth Avenue Railroad Company, upon paying to the said Eighth Avenue Railroad Co. the half part of the cost of that portion of their road lying between Varick street and West Broadway, and of keeping it in repair from time to time hereafter, and also the half of the costs and of the repairs from time to time of the extensions above authorized, shall be at liberty to use and own half of the same, and run their cars thereon, and to charge every passenger, who may come to ride any distance upon any part of their road below 43d street, the like sum of 5 cents for riding on that part of the road, and anything in the resolutions or agreement under which they are now acting, inconsistent with the said privileges, is hereby modified, so as to conform to this resolution.

Adopted by the Board of Assistants, Nov. 11, 1852.

Adopted by the Board of Aldermen, Nov. 22, 1852.

Received from his Honor the Mayor, Dec. 13, 1852, without his approval or objections thereto; therefore, under the provisions of the amended charter, the same became adopted.

Whereas, A resolution has passed the Common Council, and which resolution was permitted to become a law by the non-action of the Mayor, authorizing the 6th and 8th Avenue Railroad Company to extend their road through College place to and across Barclay street, or through Barclay or Vesey street to Broadway, and to return through Chambers street to its present termination, and

Whereas, It was the understanding, although not specified in the resolution, that said extension should and was to be only with a single track, yet the said 8th Avenue Railroad Company is now engaged in constructing said road with a double track, which will have the effect to shut from said streets all other travel, and be a serious detriment to the interests of the citizens generally, and to the owners of property on the line especially; and as there seems to be a very serious misunderstanding as to whether there should be constructed a single or a double track, therefore, be it

Resolved, That the Street Commissioner be, and he is hereby, directed to cause all further proceedings of said Companies, in the construction of said roads through the streets mentioned in the preamble, to be stayed and suspended until the further action of the Common Council, except they build a single track only through said streets, and if that course be not adopted by said Companies, that then the said Street Commissioner be, and he is hereby,

directed to have said streets restored to their former condition without delay.

Adopted by the Board of Aldermen, Dec. 18, 1852.

Adopted by the Board of Assistants, Dec. 20, 1852.

Approved by the Mayor, Dec. 21, 1852.

Resolved, That the President, Directors and Company
of the 8th Avenue Railroad be and they are hereby directed, for the better accommodation of the public, to run
their cars daily from and to 59th street and 8th avenue,
the present terminus of the rails they have already laid,
instead of, as they now run, from and to 51st street and
8th avenue.

Adopted by the Board of Aldermen, Nov. 21, 1853.

Adopted by the Board of Assistants, Nov. 29, 1853.

Approved by the Mayor, Nov. 30, 1853.

Resolved, That the 8th Avenue Railroad Company be
and are hereby directed to run their accommodation cars
from 52d to 59th street every 10 minutes, so as to start a
car every 10 minutes from 52d street up to 59th street,
and a car every 10 minutes from 59th street down to 52d
street, daily from 5 o'clock, A. M., to 10 o'clock, P. M.

Adopted by the Board of Councilmen, July 6, 1854.

Adopted by the Board of Aldermen, July 10, 1854.

Approved by the Mayor, July 12, 1854.

Resolved, That the 8th Avenue Railroad Company be
and they are hereby directed to relay the track of said

railroad, and to run their cars from 51st street to the junction of 8th avenue and Broadway, and that said Company run their cars from 51st street to 59th street every five minutes, and the Street Commissioner is hereby authorized and directed to carry this resolution into effect.

Adopted by the Board of Aldermen, November 8, 1855.

Adopted by the Board of Councilmen, December 19, 1855.

Approved by the Mayor, December 20, 1855.

Resolved, That the Street Commissioner be and is hereby directed to give the 8th Avenue Railroad Company notice to repair the pavement on the 8th avenue, in accordance with their agreement with the Corporation.

Adopted by the Board of Councilmen, May 14, 1856.

Adopted by the Board of Aldermen, May 16, 1856.

Approved by the Mayor, May 17, 1856.

Resolved, That the curb and gutter-stones on the south side of Vesey street, between Church street and Broadway, be set in 5 feet 6 inches, under the direction of the Street Commissioner, provided the expense of the same is paid by the 8th Avenue Railroad Company, and that they be directed to lay their rails as near said curb as can be done with safety to the travel for other vehicles.

Adopted by the Board of Aldermen, May 21, 1857.

Adopted by the Board of Councilmen, June 15, 1857.

Approved by the Mayor, June 19 1857.

The President of the Croton Aqueduct Board directed to notify the 8th Avenue Railroad Company to put in good repair forthwith, all the pavements in and about their rails, and, in case of neglect or refusal, to have the same done at the expense of the 8th Avenue Railroad Company.

Adopted by the Board of Aldermen, September 10, 1857.

Adopted by the Board of Councilmen, September 14, 1857.

Approved by the Mayor, September 16, 1857.

Resolved, That the 8th Avenue Railroad Company be and they are hereby required to run all their cars to 59th street regularly.

Adopted by the Board of Councilmen, February 26, 1858.

Adopted by the Board of Aldermen, March 1, 1858.

Approved by the Mayor, March 2, 1858.

Resolved, That Hudson street, from the Eighth avenue to Canal street, be paved with Belgian pavement; the Eighth Avenue Railroad Company to pay their proportion of the expense, as compelled by their grant, under the direction of the Croton Aqueduct Department.

Adopted by the Board of Councilmen, June 26, 1859.

Adopted by the Board of Aldermen, July 7, 1859.

Board of Councilmen, July 18, 1859, received from his Honor the Mayor with his objections thereto.

20

Board of Councilmen, August 8, 1859, taken up and adopted, notwithstanding the objections of his Honor the Mayor, two-thirds of all the members elected having voted therefor.

Board of Aldermen, September 5, 1859, taken up and the above action of the Board of Councilmen concurred in, two-thirds of all the members elected having voted therefor; therefore, under the provisions of the amended charter, the same became adopted.

AN ORDINANCE

IN RELATION TO THE PAVEMENT WITHIN THE TRACKS OF THE EIGHTH AND NINTH AVENUE RAILROADS.

Be it ordained by the Mayor, Aldermen and Commonalty of the City of New York, in Common Council convened:

§ 1. The Eighth Avenue Railroad Company and the Ninth Avenue Railroad Company are hereby authorized to pave all the space within the outside rails of their respective tracks with small cobble-stone, and are hereby further required to keep the said space in repair and good traveling condition.

§ 2. All resolutions, ordinances or parts thereof, so far as they conflict herewith, are hereby repealed.

§ 3. This ordinance shall take effect immediately.

Adopted by the Board of Councilmen, Sept. 8, 1859.

Adopted by the Board of Aldermen, Sept. 8, 1859.

Approved by the Mayor, Sept. 10, 1859.

Resolved, That the Eighth Avenue Railroad Company be, and they are hereby permitted to lay a turnout or switch on the south side of their tracks in Canal street, at or near Broadway, and use the same for the accommodation of their cars or those of the Ninth Avenue Railroad Company.

Adopted by the Board of Aldermen, Sept. 8, 1859.

Adopted by the Board of Councilmen, Sept. 8, 1859.

Board of Aldermen, Sept. 19, 1859, received from his Honor the Mayor with his objections thereto.

Board of Aldermen, October 3, 1859, taken up and adopted, notwithstanding the objections of his Honor the Mayor, two-thirds of all the members elected having voted therefor.

Board of Councilmen, October 3, 1859, taken up and the above action of the Board of Aldermen concurred in, two-thirds of all the members elected having voted therefor; therefore, under the provisions of the amended charter, the same became adopted.

Resolved, That the Eighth Avenue Railroad Company and the Ninth Avenue Railroad Company be and are hereby authorized and permitted in hereafter relaying their track, and in extending the same, to lay down and use the "Hewitt Bridge patent rail" or such other form of rail as may be approved of by the Street Commissioner.

Resolved, That permission is hereby granted to the Ninth Avenue Railroad Company to lay a track from the Ninth

Avenue Railroad track at Greenwich street, through Canal street, to connect with the Eighth Avenue Railroad track at Hudson street, and also to lay a track from the Ninth avenue, through Fifty-fourth street, to connect with the Eighth Avenue Railroad; and that the Eighth and Ninth Avenue Railroad Companies are hereby permitted to agree to run their cars over each other's tracks during such times as they may be respectively relaying and repairing their tracks.

Adopted by the Board of Councilmen, September 8, 1859.

Adopted by the Board of Aldermen, September 19, 1859.

Approved by the Mayor, October 3, 1859.

Resolved, That the Eighth Avenue Railroad Company is hereby authorized and directed to pave the sidewalks on the streets in front of their depot buildings, car houses, stables and workshops with Belgian pavement, and to keep such pavement at all times hereafter in good repair.

Adopted by the Board of Councilmen, November 14, 1859.

Adopted by the Board of Aldermen, December 31, 1859.

Approved by the Mayor, December 31, 1859.

NINTH AVENUE RAILROAD.

Resolved, That the Common Council do hereby grant the right and privilege to James Murphy, William Radford and Miner C. Story and their respective assigns, and to those they may associate with them, to construct a railroad from 51st street to the Battery and back, in and through the following streets, viz. :

With a double track from 51st street through the 9th avenue to Gansevoort street, thence by a single track through Greenwich street to the Battery, and by a single track through Gansevoort street to Washington street and through Washington street to the Battery, and through Battery place, between Greenwich and Washington streets, to connect the said single tracks, and also to run cars for the conveyance of passengers, &c., upon said road each and every day, at such times as they may think proper, subject to provisions hereinafter named.

Provided, Said railroad shall be constructed in all respects after the manner of the construction of the Eighth Avenue Railroad.

Provided, That in no case steam power be used on any part of said railroad ; and also

Provided, That the said grantees shall begin the construction of said railroad on or before the first day of May next, and shall complete the same and commence running cars thereon within 18 months thereafter ; and also

Provided, That the said grantees shall run cars upon the road so constructed each way, between 51st street and the

Battery, every day, as often as 15 minutes from 5 to 6 A. M., and every 4 minutes from 6 A. M. to 12 M., every 15 minutes from 8 A. M. to 12 M., and as much oftener as public convenience may require, under such directions as the Common Council may, from time to time, prescribe; and

Provided, That no more than five cents be charged for each passenger riding over the whole or any portion of the distance of said road; and also

Provided, That said grantees shall keep the space, for two feet each side of the same, at all times in thorough repair, and also

Provided, That the said **cars** shall be licensed by the Mayor, and the grantees shall pay the annual fee of $20 per car for such license : and

Also, the said grantees and their associates and assigns, shall have the privilege to organize a joint-stock association, either with or without incorporation, to carry out the objects of this grant, and a majority in interest of the grantees, their assigns and associates, shall have the control, management and direction of the road and the business thereof; and should any or either of the grantees or their associates, or of the shareholders, neglect to pay their respective proportion of the money required for carrying into full effect the grant hereby made, when by such majority thereunto required, the others shall be at liberty to make such payment; and this grant shall enure to the benefit of those who pay in the proportion of their respective contributions.

These resolutions shall be certified by said grantees above named; and a copy thereof, signed by them, shall be deemed the agreement between the Mayor, Aldermen and Commonalty of the city of New York and said associates ; and shall be sufficient in all respects to give and grant to the said grantees, their associates and assigns, aforesaid, the right and privilege above mentioned, and bind them to conform to the directions herein contained.

And also provided, that said railroad shall be continued from Fifty-first street along the Ninth avenue to the Bloomingdale road, to the Tenth avenue, thence along the Tenth avenue to the Harlem river, whenever required by the Common Council and as soon and as fast as said avenues are graded.

Adopted by the Board of Assistants, Dec. 20, 1852.

Adopted by the Board of Aldermen, Jan. 5, 1853.

Received from his Honor the Mayor, Jan. 12, 1853, with his objections thereto.

Board of Aldermen, November 14, 1853, taken up, reconsidered and adopted, notwithstanding the objections of his Honor the Mayor thereto, two-thirds of all the members elected voting in favor thereof.

Board of Assistants, December 28, 1853, taken up, reconsidered and adopted, notwithstanding the objections of his Honor the Mayor thereto, two-thirds of all the members elected voting in favor thereof; therefore, under the provisions of the amended charter the same became adopted.

D. T. VALENTINE,
Clerk of Common Council.

JAMES MURPHY,
WILLIAM RADFORD,
MINER C. STORY.

CITY AND COUNTY OF NEW YORK, *ss.*:—We, James Murphy, William Radford, and Miner C. Story the grantees in the foregoing resolutions do hereby certify as therein required that the said resolutions, and the conditions and provisions therein specified, contain the agreement between us and the Mayor, Aldermen and Commonalty of the city of New York, in relation to the right, privilege and grant made and conferred upon us and our associates and assigns, by said Mayor, Aldermen and Commonalty, in and by the same.

In witness whereof, we have hereunto subscribed our names at the city of New York, the thirtieth day of December, one thousand eight hundred and fifty-three.

JAMES MURPHY,
WILLIAM RADFORD
MINER C. STORY.

In the presence of
    JOHN B. HASKIN.

Resolved, That Messrs. Story, Radford & Murphy, their associates and assigns, to whom was granted the right in 1853 to build a railroad in Ninth avenue, Gansevoort, Washington and Greenwich streets and Battery place, be and are hereby directed to proceed, immediately upon the removal of the injunction issued against them by the Supreme Court by which they have been prevented from completing their road, with the construction of said road and complete the same from Fifty-first street to Battery place within six months thereafter, in the manner set forth in said grant.

Resolved, That, until the said injunction be removed, the said grantees be directed to place and run cars upon such portion of said road as is completed and not affected by the said injunction, thereby granting to the public all the accommodation that the circumstances will allow, the said grantees being permitted, in consideration therefor, to make a charge of five cents as fare for each passenger carried by them over such portions of their road.

Adopted by the Board of Councilmen, April 29, 1858.

Adopted by the Board of Aldermen, May 13, 1858.

Approved by the Mayor, May 24, 1858.

Resolved, That the Street Commissioner be, and he is hereby, authorized and directed, forthwith, to cause the rails of the Ninth Avenue Railroad Company now laid at the intersection of Washington and Bethune streets, to be taken up, the bridges removed, the filth taken away, and the intersections repaved in such a manner that the surface water and filth will flow towards the North River without obstruction, and that the expense thereof be charged to the Ninth Avenue Railroad Company.

Adopted by the Board of Aldermen, Oct. 7, 1858.

Adopted by the Board of Councilmen, Oct. 11, 1858.

Approved by the Mayor, Oct. 21, 1858.

Whereas, The Corporation of the city of New York granted to James Murphy and others the right and privilege or constructing and operating a railroad in Ninth avenue and other streets, and inasmuch as the grantees have laid the rails of their said railroad from Fifty-fourth street to Canal street but have been prevented by legal difficulties, interposed by property-owners below Canal street, from completing their railroad, and

Whereas, The necessary accommodation of the public requires that the said grantees should be allowed an outlet or terminus for their road in the southern business portion of the city, and that they should be compelled to put their railroad into active operation ; therefore be it

Resolved, That James Murphy and others, the owners of the said railroad grant or license, and their assigns, are hereby authorized and directed to connect their railroad track with the tracks of the Hudson River Railroad and Sixth and Eighth Avenue Railroad Companies, in and below Canal street, and run their cars upon any portion of the tracks of the said other railroads, in and below Canal street, and to charge the rate of fare provided in their said grant or license.

Adopted by the Board of Councilmen, June 24, 1859.

Adopted by the Board of Aldermen, June 27, 1859.

Approved by the Mayor, July 2, 1859.

AN ORDINANCE

IN RELATION TO THE PAVEMENT WITHIN THE TRACKS OF THE EIGHTH AND NINTH AVENUE RAILROADS.

Be it ordained by the Mayor, Aldermen, and Commonalty of the City of New York, in Common Council convened:

§ 1. The Eighth Avenue Railroad Company and the Ninth Avenue Railroad Company are hereby authorized to pave all the space within the outside rails of their respective tracks with small cobble-stone, and are hereby further required to keep the said space in repair and good traveling condition.

§ 2. All resolutions, ordinances or parts thereof, so far as they conflict herewith, are hereby repealed.

§ 3. This ordinance shall take effect immediately.

Adopted by the Board of Councilmen, September 8, 1859.

Adopted by the Board of Aldermen, September 8, 1859.

Approved by the Mayor, September 10, 1859.

Resolved, That the Eighth Avenue Railroad Company and the Ninth Avenue Railroad Company be and are hereby authorized and permitted, in hereafter relaying their track and in extending the same, to lay down and use the " Hewitt Bridge Patent Rail, or such other form of rail, as may be approved of by the Street Commissioner.

Resolved, That permission is hereby granted to the Ninth

Avenue Railroad Company to lay a track from the Ninth Avenue Railroad track at Greenwich street, through Canal street, to connect with the Eighth Avenue Railroad track at Hudson street, and also to lay a track from the Ninth avenue, through Fifty-fourth street, to connect with the Eighth Avenue Railroad; and that the Eighth and Ninth Avenue Railroad Companies are hereby permitted to agree to run their cars over each other's tracks during such times as they may be respectively relaying and repairing their tracks.

Adopted by the Board of Councilmen, September 8, 1859.

Adopted by the Board of Aldermen, September 19, 1859.

Approved by the Mayor, October 3, 1859.

Resolved, That the Ninth Avenue Railroad Company is hereby authorized and directed to pave the sidewalks on the streets in front of their depot buildings, car houses, stables and work shops with Belgian pavement, and to keep such pavement, at all times hereafter, in good repair.

Adopted by the Board of Councilmen, November 14, 1859.

Adopted by the Board of Aldermen, December 31, 1859.

Approved by the Mayor, December 31, 1859.

HUDSON RIVER RAILROAD.

AN ORDINANCE.

The Mayor, Aldermen and Commonalty of the City of New York, in Common Council convened, do ordain as follows:

SECTION 1. Permission is hereby granted to the Hudson River Railroad Company to construct a double track of rails, with suitable turn-outs, along the line of the Hudson River, from Spuyten Devil Creek to near 68th street; occupying so much of the 12th avenue as lies along the shore, thence winding from the shore so as to intersect the 11th avenue at or near 60th street; thence through the middle of the 11th avenue to about 32d street; thence on a curve across to the 10th avenue, intersecting the 10th avenue at or near 30th street; thence through the middle of the 10th avenue to West street, and thence through the middle of West street to Canal street.

§ 2. The said Hudson River Railroad Company shall grade, pave, and keep in repair a space 25 feet in width, in and about the tracks, in all the avenues and streets through which the said track or tracks shall be laid, whenever the Common Council shall deem the interest of the public to require such pavement to be done. The said Company shall lay such rail track through the avenues and streets in conformity to such directions as to line and grade as shall be given by the Street Commissioner, and shall conform their said railroad to the grades of the avenues and streets through which it shall extend or cross, as shall be, from time to time, established by the Common Council, if the latter so require; and shall lay their rails or tracks in the streets or avenues in such manner as to cause no unnecessary impediment to the common or ordinary use of the street for all other purposes, and so as to leave all the water courses free and unobstructed. It shall be especially incumbent on the said Hudson River Railroad Company, at their own cost, to construct stone bridges across such of the streets intersected by the railroad as may, by the elevation of their grades above the surface of the said road, require to be arched or bridged, whenever, in the opinion of the Common Council, the same shall be necessary for public convenience; and also to make such embankments or excavations, as the Common Council may deem necessary to render the passage over the railroad and embarkments at the cross streets easy and convenient for all the purposes for which streets and roads are usually put to, and the said Company shall also make, at their own cost and charge, all such drains and sewers as their embankments or excavations may, in the opinion of the Common Council, render necessary; and said Company shall be at all times subject to such regulations, with reference to the convenience of public travel through such streets and avenues as are affected by the said railroad, as the Common Council shall, from time to time by resolution or ordinance, direct; and the Corporation hereby reserves the right to require said Company, at any time after the 11th avenue shall be made to 14th street, to take up their rails in the 10th avenue and lay them in the 11th avenue to said 14th street, and through 14th street to connect with West street.

§ 3. The said Company shall, within one year from the passage of this ordinance, and before entering upon any contracts for grading, file in the office of the Street Commissioner a map showing the location and intended grade of said railroad.

§ 4. Permission is hereby granted to the Hudson River Railroad Company to run their locomotives as far south as 30th street, and no further.

§ 5. The said Hudson Railroad Company shall be and are hereby prohibited from running a stated train between any points below 32d street for the carrying of passengers between those points, under the penalty of $25 for each passenger from whom fare shall be received therefor.

§ 6. This ordinance shall not be construed as binding upon the Corporation, nor shall it go into effect until the said Hudson River Railroad Company shall first duly execute, under their corporate seal, such an instrument in writing, covenanting and engaging, on their part and behalf, to stand to, abide by and perform all such conditions and requirements contained in the 2d and 3d sections of the ordinance, as the Mayor and the Counsel to the Corporation shall by their certificate approve, and not until such instrument shall be filed, so certified, in the office of the Comptroller of this city.

Adopted by the Board of Aldermen, April 30, 1847.

Adopted by the Board of Assistants, May 3, 1847.

Approved by the Mayor, May 6, 1847.

To all to whom these presents shall come, greeting :

Whereas, The Mayor, Aldermen and Commonalty of the city of New York, by an ordinance approved on the sixth day of May, A. D. 1847, gave consent to the Hudson River Railroad Company to commence in the city of New York and construct therein a double track of rails, with suitable turnouts along the line therein mentioned, from Canal street to the Spuytenduyvel Creek, and did, in and by said ordinance, assent to the location by the directors of said company of said railroad on and over the streets and avenues mentioned in said ordinance and crossed by said line ; and

Whereas, Pursuant to said ordinances and the acts incorporating said company and amendatory thereof, the said directors have located the said railroad in the city of New York, according to the map prepared to be filed herewith, showing the location and intended grade of the Hudson River Railroad in the city of New York,

Now, know ye that the said the Hudson River Railroad Company, for themselves and their successors, do hereby, in the consideration of the premises, covenant and engage, to and with the Mayor, Aldermen and Commonalty of the city of New York, and their successors forever, to grade, regulate, pave and keep in repair a space twenty-five feet in width, in and about the tracks in all the avenues an streets through which the said track or tracks shall be laid, whenever the Common Council shall deem the interest ot the public to require such pavement to be done.

And, that the said company will lay such rail track

through the avenues and streets in conformity to such direction, as to line and grade, as shall be given by the Street Commissioner, and shall conform their said railroad to the grades of the avenues and streets through which it shall extend or which it shall cross, as shall be, from time to time, established by the Common Council, if the latter so require.

And that said company will lay their rails, or tracks, in the streets or avenues in such manner as to cause no unnecessary impediment to the common and ordinary use of the streets for all other purposes, and so as to leave all the water courses free and unobstructed.

And further, that said company will at their own cost construct stone bridges across such of the streets, intersected by the said railroad, as may, by the elevation of their grades above the surface of said road, require to be arched or bridged, whenever, in the opinion of the Common Council, the same shall be necessary for public convenience.

And also, that the said company will make such embankments or excavations as the Common Council may deem necessary, to render the passage over the said railroad and embankments at the cross-streets easy and convenient for all purposes to which streets and roads are usually put.

And, that the said company will also make, at their own cost and charge, all such drains and sewers as their embankments or excavations may, in the opinion of the Common Council, render necessary.

And will at all times be subject to such regulations, with

reference to the convenience of public travel through such streets and avenues, as are affected by said railroad, as the Common Council shall, from time to time, by resolution or ordinance direct.

And further, that, if thereto required by the Corporation at any time after the Eleventh avenue shall be made to Fourteenth street, the said Company will take up their rails in the Tenth avenue and lay them in the Eleventh avenue to said Fourteenth street, and through Fourteenth street to connect with West street.

And, that the said company will, within one year from the passage of the said ordinance, and before entering upon any contracts for grading, file, in the office of the Street Commissioner, a map showing the location and intended grade of said railroad.

And, lastly, that said Company will stand to, abide by and perform, all and singular, the conditions and requirements contained in the second and third sections of the said ordinance.

In witness whereof, the said the Hudson River Railroad Company have hereunto affixed the corporate seal this 12th day of August, A. D. 1847.

WM. CHAMBERLAIN, [L. s.]

President.

I, William V. Brady, Mayor of the City of New York, do hereby certify that I approve of the preceding covenant as being in compliance with the ordinance of the Corporation approved May 6, 1847, referred to in said covenant.

WM. V. BRADY,

Mayor.

I, Willis Hall, Counsel to the Corporation of the city of New York, do hereby certify that I approve of the preceding covenant, as being in compliance with the ordinance referred to in the above certificate.

WILLIS HALL,

Counsel of Corporation.

City and County of New York: On this nineteenth day of August, A. D. 1847, before me personally appeared William Chamberlain, known to me to be the President of the Hudson River Railroad Company, and, being by me duly sworn, did depose and say that he resides in the city of New York, that the seal thereto affixed is the seal of the said company, and that the same was so affixed by their authority.

JOSEPH STRONG,

Commissioner of Deeds.

Resolved, That the Hudson River Railroad Company be authorized to lay down a double track of rails, with suitable curves and turn-out, from the northerly line of Canal street, at West street, through Canal and Hudson streets to Chambers street, under the direction of the Street Commissioner, and subject to all the restrictions, obligations, provisions and conditions of the ordinance authorizing said Company to lay down rails to Canal street.

Adopted by the Board of Aldermen, Aug. 1, 1849.

Adopted by the Board of Assistants, Sept. 24, 1849.

Approved by the Mayor, Sept. 25, 1849.

Resolved, That the Hudson River Railroad Company may extend one of their tracks around the country market (leased to them at foot of Canal street), with suitable curves and turn-outs, under the direction of the Street Commissioner, so as to connect with the track on West and Canal streets already constructed by them, subject to all the terms, conditions and restrictions of the annexed resolution passed and approved as stated below. (See resolution approved Sept. 25, 1849.)

Adopted by the Board of Aldermen, Dec. 24, 1849.

Adopted by the Board of Assistants, Dec. 28, 1849.

Approved by the Mayor, Jan. 7 (10 A. M.), 1850.

Resolved, That the market-house, and block of ground on which it stands, bounded by Washington and West streets, and Canal and Hoboken streets, be leased to the Hudson River Railroad Company for a passenger depot, for the term of ten years from the 1st of May, 1849, at the rent of $2,000 per annum, payable quarterly, subject to renewal for a further term of ten years at a rent to be determined by two appraisers mutually chosen, with power to select a third in case they cannot agree; said appraisers to be duly sworn before entering upon their duties.

Adopted by the Board of Assistants, April 23, 1849.

Adopted by the Board of Aldermen, April 30, 1849.

Approved by the Mayor, May 3, 1849.

Resolved, That the Hudson River Railroad Company have permission to run their dumb engine to Chambers street, to test its power and probable safety for conducting their cars to Chambers street, under the direction of the Street Commissioner.

Adopted by the Board of Aldermen, July 6, 1850.

Adopted by the Board of Assistants, July 8, 1850

Approved by the Mayor, July 9, 1850.

Resolved, That the Hudson River Railroad Company be directed to build a substantial addition to the pier at Manhattanville, foot of 130th street, North River, by extending the same into the river for a distance equivalent to the portion of said pier cut off between the railway and the shore (the same being in length about 215 feet), the addition to consist of blocks and bridges, under such directions as shall be given by the Street Commissioner, and the work to be commenced on or before the 1st day of April, 1851, and completed within three months thereafter. And in case the Hudson River Railroad Company shall neglect to comply with this resolution and commence to build and complete the said pier at or before the periods herein mentioned, that then the said addition shall be built by the Street Commissioner, and legal measures taken by the Counsel to the Corporation to compel payment, by the Hudson River Railroad Company, of the expenses incurred : and the sum of $8,000 is hereby authorized to be taken from the appropriation for Docks and Slips, to carry this resolution into effect.

Resolved, That the Street Commissioner be directed to notify the Hudson River Railroad Company of the passage of the foregoing resolution.

Adopted by the Board of Assistants January 22, 1851.

Adopted by the Board of Aldermen, January 30, 1851.

Approved by the Mayor, February 10, 1851.

Resolved, That the Comptroller be and he is hereby directed to prepare and cause to be executed to the Hudson River Railroad Company, a lease of the piece of ground described in the annexed map, and bounded by 12th, Washington, Gansevoort, West street and Tenth avenue, for a term of fifteen years from the 1st of May, 1851, at a rent of $6,192 per annum, payable quarterly, subject, however, to the present existing leases to F. Depeyster, Foster & Van Nostrand, Patrick Noonan, and Edward W. Phelps ; said lease to contain a covenant for two renewals for 15 years each, at a rent to be fixed by appraisement at the expiration of each term ; also a covenant that the said Company pay all taxes imposed on said property during said terms ; and also all assessments, ordinary or extraordinary, except the assessment that may be imposed for the continuation of Washington to 12th street ; said ground to be occupied exclusively by said Company as a depot and other purposes connected with the said road.

Adopted by the Board of Aldermen, May 30, 1851.

Adopted by the Board of Assistants, June 2, 1851.

Approved by the Mayor, June 4, 1851.

Resolved, That the offer of the Hudson River Railroad Company, contained in the annexed petition, be accepted, and that the said sum of ten thousand dollars be paid by said Company on the contract for the bulkhead between the present pier, at 130th street, to the north line of 131st street, and upon said Company releasing all right to land under water conveyed to them by Schiefflin & Lawrence, south of the north line of 131st street, and upon such release being executed the Corporation to release the Company from the covenants in the grants to Schiefflin & Lawrence relating to the land so released.

Resolved, That the Street Commissioner be directed to proceed and build a bulkhead from the pier at the foot of 130th street to the north line of 131st street, on the North River, and a pier at the foot of 131st street, and that said Street Commissioner proceed forthwith to build and erect the same ; and that the said sum of ten thousand dollars, to be received from the Hudson River Railroad Company, be appropriated and applied for that purpose ; said bulkhead and pier to be completed within one year.

Resolved That a grant, confirming to said Railroad Company their roadway as now constructed, 78 feet in width, between 130th and 132d streets, and according to the map filed with the Clerk of the Common Council, and release the said Company from all claim upon them for building or extending any pier or bulkhead opposite the premises in question.

Resolved, That the resolution approved by the Mayor January 5, 1850, leasing to Francis R. Tillou the dock at the foot of 130th street for a public ferry, at the rent of $50 per annum, be repealed.

Adopted by the Board of Aldermen, June 4, 1851.

Adopted by the Board of Assistants, June 4, 1851.

Approved by the Mayor, June 5, 1851.

Resolved, That the Hudson River Railroad Company be,

and they are hereby, directed to take up their rails on the Tenth avenue, from Thirtieth street, fifty feet south, and relay them in a direct line until they reach Thirtieth street, before they commence making their curve towards the Eleventh avenue.

Adopted by the Board of Aldermen, May 29, 1851.

Adopted by the Board of Assistants, June 2, 1851.

Received from his Honor the Mayor, August 8, 1851, without his approval or objections thereto; therefore, under the provisions of the amended charter, the same became adopted.

Resolved, That the Hudson River Railroad Company be directed to take up their rails, and relay them so that at the southwest corner of Tenth avenue and Thirtieth street they shall be distant from the angle of the curb at least twelve feet.

Adopted by the Board of Assistants, August 6, 1851.

Adopted by the Board of Aldermen, August 7, 1851.

Approved by the Mayor, August 11, 1851

Whereas, By certain resolutions of the Common Council, approved by the Mayor, June 5, 1851, certain claims, existing between the Corporation and the Hudson River Railroad Company, in relation to the dock, &c., at Manhattanville were adjusted, by which adjustment the said Company were to pay $10,000, and assign certain water grants, owned by them, to the City; and whereas, said Company, on account of their large expenditure at the present time consequent upon the completion of their road, desire to substitute a bond of the Company, payable at a future day, with interest thereon, in lieu of the present payment of the said amount, therefore

Resolved, That the Comptroller be and he is hereby authorized to receive a bond of the Hudson River Railroad Company for $10,000, with interest, at six per cent. per annum, payable quarterly, and the principal sum payable in five years, in lieu of the present payment of said $10,000, as provided for in said resolutions.

Adopted by the Board of Aldermen, Nov. 11, 1851.

Adopted by the Board of Assistants, Nov. 17, 1851.

Approved by the Mayor, November 19, 1851.

Resolved, That the Hudson River Railroad Company be required to take up the present rails in Hudson street, as also in Canal and West streets, and put down a grooved rail, similar to the one in Park row put down by the Harlem Railroad Company, and that the Hudson River Railroad Company be required to complete the same in eight months from the passage of this resolution; and in case of their failing to comply herein, then the Street Commissioner be, and he is hereby required to take up the rails of said track, from Chambers to Thirty-first street, and repair said street in like manner as previous to the occupancy of said street by said Railroad Company.

Adopted by the Board of Assistants, Oct. 26, 1852.

Adopted by the Board of Aldermen, Nov. 8, 1852.

Received from his Honor the Mayor, December 4, 1852, without his approval or objection thereto; therefore, under the provisions of the amended charter, the same became adopted.

Resolved, That permission be, and is hereby, given to the Hudson River Railroad Company to lay grooved rails for a track on Pier No. 48, North River, foot of Clarkson street, for and during the continuance of the lease of sadt pier, the said track to connect with their railroad in West street.

Adopted by the Board of Aldermen, Dec. 27, 1853.

Adopted by the Board of Assistants, Dec. 29, 1852.

Approved by the Mayor December 30, 1852.

Resolved, That the Street Commissioner be, and he is hereby directed to notify the Hudson River Railroad Company to remove, and hereafter to refrain from standing the cars of their Company in Hudson street; and, further, if said Company refuse to comply therein, then the Street Commissioner is hereby directed to remove the same in accordance with the ordinances of the Corporation.

Adopted by the Board of Councilmen, March 13, 1854.

Adopted by the Board of Aldermen, June 10, 1854.

Approved by the Mayor, July 13, 1854.

Resolved, That the Hudson River Railroad Company be directed to cause Hudson street, from Canal to Chambers street, to be repaired, according to the terms of their grant.

Adopted by the Board of Councilmen, January 19, 1855.

Adopted by the Board of Aldermen, February 11, 1855
Approved by the Mayor, February 16, 1855.

Resolved, That the Hudson River Railroad Company be, and are hereby notified to fill in the low grounds adjoining their property, between 12th avenue and Hudson River, and 130th and 131st streets ; and that unless the said Company forthwith comply with this resolution, that the Street Commissioner be, and is hereby directed to cause a notice to be sent to said Railroad Company, to the effect that, in their default, for one month after notice, he will cause the said filling to be done at the expense of said Railroad Company.

Adopted by the Board of Councilmen, October 22, 1855.

Adopted by the Board of Aldermen, August 11, 1856.

Approved by the Mayor, August 13, 1856.

The President of the Croton Aqueduct Board directed to notify the Hudson River Railroad Company to put in good repair, forthwith, all the pavements in and about their rails, and, in case of neglect or refusal, to have the same done at the expense of the Hudson River Railroad Company.

Adopted by the Board of Aldermen, September 10, 1857.

Adopted by the Board of Councilmen, September 14, 1857.

Approved by the Mayor, September 16, 1857.

Resolved, That the Hudson River Railroad Company be, and are hereby required to remove the present high rail in use upon their road, from the corner of Chambers street

and West Broadway up to 53d street, and to lay down in
the stead thereof the rail known as the grooved rail, and
that the same be done within six months from the passage
of this resolution by the Common Council.

Resolved, That the Hudson River Railroad Company be,
and they are hereby, authorized and directed to place upon
their road city passenger or small cars, to be run between
the depot at Chambers street and 53d street; to take up
and set down city passengers between those points; to be
governed by the general rules regulating the Eighth Avenue Railroad; and, further, that they run a car thereon
each and every day, both ways, as often as every 15
minutes from 5 to 6 o'clock, A. M., and every 5 minutes
from 6 o'clock, A. M., to 8 o'clock, P. M.; every 15 minutes
from 8 o'clock, P. M., to 12 o'clock, P. M., and every 30
minutes from 12 o'clock, P. M., to 5 o'clock, A. M., and as
much oftener as public convenience may require, under
the regulations of the Common Council; and that the said
Company shall have the right to demand and receive from
each passenger conveyed in said cars the sum of five (5)
cents, and no more. The aforesaid cars to be placed and
run upon said road within six months from the passage of
this resolution by the Common Council. It being a spe-
cial permission and understanding in making this grant to
the Hudson River Railroad Company, that the said Company shall not, at any time, either directly or indirectly, in
any way alienate from themselves. as a Company, or in
any manner dispose of the right to run small cars upon
their said road, hereby granted, unless by consent of the
Common Council, under the penalty of the forfeiture of
this grant immediately thereupon.

Resolved, That the Hudson River Railroad Company be, and they are hereby directed to cease the running of locomotives or steam engines below Fifty-third street immediately upon the small cars being placed upon their road, in accordance with the foregoing resolution.

Adopted by the Board of Aldermen, November 22, 1858.

Adopted by the Board of Councilmen, December 2, 1858.

Approved by the Mayor, December 13, 1858.

Resolved, That the Hudson River Railroad Company be, and they are hereby directed to have the grade of their railroad conform to the grade of Eleventh avenue, as heretofore put under contract, in order that the grading of said avenue may be completed, under the direction of the Street Commissioner.

Adopted by the Board of Councilmen, April 11th, 1859.

Adopted by the Board of Aldermen, June 30th, 1859.

Approved by the Mayor, July 2d, 1859.

Resolved, That at the expiration of the lease now held by the Hudson River Railroad Company, of the building on the plot of ground bounded by Canal, West and Hoboken streets, the said building be removed, under the direction of the Street Commissioner, and that the vacant space be appropriated as a country market.

Adopted by the Board of Aldermen, Dec. 19, 1859.

Adopted by the Board of Councilmen, Dec. 31, 1859.

Approved by the Mayor, Dec. 31, 1859.

RAILROAD IN FORTY-SECOND STREET.

Resolved, That permission be, and is hereby given to A. M. Allerton, Jr. and Company, to lay a track for a railroad on 42d street, to connect with the Hudson River Railroad at the 11th avenue and to extend to the Hudson River, to be used to convey stock, such as cattle, sheep and hogs ; the said track to remain during the pleasure of the Common Council, the Common Council to give 60 days' notice for its removal.

Adopted by the Board of Councilmen, Dec. 20, 1858.

Adopted by the Board of Aldermen, Dec. 30, 1858.

Approved by the Mayor, Jan. 8, 1859.

ORDINANCES

AFFECTING THE

CITY RAILROADS,

OR

GENERAL IN THEIR APPLICATION.

---•·•---

AN ORDINANCE.

Be it ordained by the Mayor, Aldermen, and Commonalty of the City of New York, in Common Council convened:

That the grantees of all railroads within the city, their associates and successors, shall, in the construction, alteration and repairs of such railroads, at all times, furnish such new work, make such additions, and do all such repairs to man-hole heads and covers, receiving-basins, and stop-cocks and covers, and generally of all fixtures connected with sewers and the distribution of Croton Water, as may, in the process of laying down such rail tracks, be affected thereby; such additions, alterations and repairs to be done under the direction of, and to the satisfaction of the Croton Aqueduct Department, and that in no case shall such rail tracks be laid over the line of Croton water mains, stop-cocks or sewer man-holes.

Adopted by the Board of Aldermen, April 25, 1853.

Adopted by the Board of Assistants, May 13, 1853.

Approved by the Mayor, May 16, 1853.

Resolved, That the Common Council of the city of New York disapprove of the Legislature passing any law granting the privilege of railroads in the city of New York.

Resolved, That the above resolution be duly authenticated, and immediately sent to the Board of Councilmen, and if approved, that it be sent to the Legislature, through some of our Senators or Representatives, for their action.

Adopted by the Board of Aldermen, April 5, 1854.

Adopted by the Board of Councilmen, April 5, 1854.

Approved by the Mayor, April 6, 1854.

Resolved, That the President of the Croton Aqueduct Board be, and he is hereby, instructed to notify the several Railroad Companies to put in good repair, forthwith, all the pavements in and about their rails, in accordance with their agreements; and in case all or any of said Companies (viz., the Harlem, Hudson, Sixth, Eighth, Second, and Third Avenue Railroad Companies) refuse or neglect to comply with said notice, that the said President of the Croton Aqueduct Board cause the same to be done at the expense of the companies interested.

Adopted by the Board of Aldermen, Sept. 10, 1857.

Adopted by the Board of Councilmen, Sept. 14, 1857.

Approved by the Mayor, Sept. 16, 1857.

Resolved, That the Croton Aqueduct Board be, and they are hereby instructed to notify the several railroad companies, to wit: The New York and Harlem, Hudson River, Sixth Avenue, Eighth Avenue, Second Avenue, and Third Avenue Railroad Companies to put in good repair, within twenty days from the date of the service of a notice from said Board to that effect, all the pavements in and about their respective rail tracks, in accordance with the terms and conditions of their several grants from and agreements with the Mayor, Aldermen and Commonalty of this city, and in case said Companies, or either of them, shall neglect or refuse so to do, that said Board shall be, and they are hereby authorized and directed to cause the same to be done at the expense of the company or companies so neglecting or refusing to comply with said notice, and the expense incurred by said Board in so doing shall be certified to by them, and placed in the hands of the Corporation Counsel for immediate collection by process of law; and that the sum of $5,000 be, and it is hereby specially appropriated for the purpose of enabling said Board to carry this resolution into immediate effect.

Adopted by the Board of Councilmen, April 26, 1858.

Adopted by the Board of Aldermen, May 10, 1858.

Approved by the Mayor, May 11, 1858.

AN ORDINANCE

FOR THE LICENSING OF CITY RAILROAD PASSENGER CARS

The Mayor, Aldermen and Commonalty of the City of New York, in Common Council convened, do ordain as follows ·

SECTION 1. Each and every passenger railroad car running in the city of New York below One Hundred and

Twenty-fifth street, shall pay into the city treasury the sum of 50 dollars, annually, for a license, a certificate of such payment to be procured from the Mayor, except the small one-horse passenger cars, which shall each pay the sum of 25 dollars, annually, for said license, as aforesaid.

§ 2. Each certificate of payment of license shall be affixed to some conspicuous place in the car, that it may be inspected by the proper officers.

§ 3. For every passenger car run upon any of the city railroads below One Hundred and Twenty-fifth street without the proper certificate of license, the proprietor or proprietors thereof shall be subject to a penalty of 50 dollars, to be recovered by the Corporation Attorney, as in the case of other penalties, and for the benefit of the city treasury.

§ 4. This ordinance shall go into effect immediately.

Adopted by the Board of Councilmen, Dec. 13, 1858.

Adopted by the Board of Aldermen, Dec. 22, 1858.

Approved by the Mayor, Dec. 31, 1858.

A message having been received from his Honor the Acting Mayor, in relation to sundry proposed railroad charters about to be granted by the Legislature, the following resolution was adopted :

Resolved, That the Counsel to the Corporation be, and he is hereby authorized and directed immediately to take all proper legal measures to restrain and prevent the use or occupation of any street, public place or highway in the city of New York, by any person or persons, company or com panies, corporation or corporations, claiming by any act or acts of the Legislature of this State, at its recent or any previous session, the right, exclusive or otherwise, of laying rails and running cars thereon in any such street, public place or highway in the said city of New York, without the consent of the Mayor, Aldermen and Commonalty of the said city being first had and obtained.

Adopted by the Board of Aldermen, April 20, 1860.

Adopted by the Board of Councilmen, April 20, 1860.

Approved by the Mayor, April 20, 1860.

STATE LAWS

RELATIVE TO THE

CITY RAILROADS.

AN ACT *relative to the construction of Railroads in Cities.*

Passed April 4, 1854.

The People of the State of New York, represented in Senate and Assembly, do enact as follows:

SECTION 1. The Common Councils of the several cities of this State shall not, hereafter, permit to be constructed, on either of the streets or avenues of said city, a railroad for the transportation of passengers, which commences and ends in said city, without the consent thereto of a majority in interest of the owners of property upon the streets in which said railroad is to be constructed, being first had and obtained. For the purpose of determining what constitutes said majority in interest, reference shall be had to the assessed value of the whole located upon such street or avenue.

§ 2. After such consent is obtained, it shall be lawful for the Common Council of the city, in which such street or avenue is located, to grant authority to construct and establish such railroad upon such terms, conditions and stipulations, in relation thereto, as such Common Council may see fit to prescribe. But no such grants shall be made except to such person or persons as shall give adequate security to comply in all respects with the terms, conditions, and stipulations, so to be prescribed by such Common Council, and will agree to carry and convey passengers upon such railroad at the lowest rates of fare. Nor shall such grants be made until after public notices of intention to make the same, and of the terms, conditions and stipulations, upon which it will be given, and inviting proposals therefor at a specified time and place, shall be published, under the direction of the Common Council, in one or more of the principal newspapers published in the city in which said railroad is proposed to be authorized and constructed.

§3. This act shall not be held to prevent the construction, extension or use of any railroad in any of the cities of this State which have already been constructed in part; but the respective parties and companies by whom such roads have been in part constructed, and their assigns, are hereby authorized to construct, complete, extend and use such roads, in and through the streets and avenues designated in the respective grants, licenses, resolutions or contracts, under which the same have been so in part constructed, and to that end the grants, licenses and resolutions aforesaid are hereby confirmed.

§4. This act shall take effect immediately.

CHAPTER 373.

AN ACT *in relation to the Second Avenue Railroad Company, of the city of New York.*

Passed April 12, 1855.

The People of the State of New York, represented in Senate and Assembly, do enact as follows :

§ 1. The Second Avenue Railroad Company of the city of New York, are hereby authorized to construct a bridge for the use of their road across the Harlem River, at and from the termination of the Second avenue in said city.

§ 2.   *   *   *   *   *   *   *

§ 3. The said bridge across the Harlem River shall be constructed with a draw of not less than sixty feet in width, in the clear, and with piers not less than sixty feet apart.   *   *   *   *

§ 4. The said bridges shall be constructed and maintained by said company, in such manner as not to unnecessarily impede or obstruct the navigation of said rivers.

§ 5. Said company shall keep some competent person stationed at each of the several draw-bridges constructed by them over said rivers, whose duty it shall be to swing the draws whenever any vessel is approaching and about to pass either of said bridges.

§ 6.   *   *   *   *   *   *   *

§ 7. This act shall take effect immediately.

CHAPTER 10.

AN ACT *relative to Railroads in the City of New York.*

Passed January 30th, 1860.

The People of the State of New York, represented in Senate and Assembly, do enact as follows :

§ 1. It shall not be lawful hereafter to lay, construct or operate any railroad in, upon or along any or either of the streets or avenues of the city of New York, wherever such railroad may commence, or end, except under the authority and subject to the regulations and restrictions which the Legislature may hereafter grant and provide. This section shall not be deemed to affect the operation, as far as laid, of any railroad now constructed and duly authorized. Nor shall it be held to impair in any manner any valid grant for or relating to any railroad in said city existing on the first day of January, eighteen hundred and sixty.

§ 2. All acts and parts of acts inconsistent with this act are hereby repealed.

§ 3. This act shall take effect immediately.

[NOTE.—The following five grants for railroads in the city of New York were passed by the last State Legislature, but have not, as yet, been acknowledged nor confirmed by the Common Council of the city.]

CHAPTER 511

AN ACT *to authorize the construction of a railroad track on South, West, and certain other streets in the city of New York.*

Passed April 17, 1860; notwithstanding the objections of the Governor.

The People of the State of New York, represented in Senate and Assembly, do enact as follows:

SECTION 1. Charles W. Durant, Myron S. Clark, John Butler, Jr., Henry Rigley, Orson H. Sheldon, Warren E. Russel, Robert B. Van Valkenburgh, John Myers, John De La Montagnie, William R. Stewart, James S. Leach, James S. Sluyter, John A. Cooke, James C. Kennedy, Thomas C. Durant, Benjamin F. Bruce, Edward R. Phelps, Augustus L. Brown, and their assigns, are hereby authorized to lay, construct, operate, and use a railroad, with a double or single track, as hereinafter provided, and to convey passengers and freight thereon for compensation, through, upon, and along the following streets and avenues, route, or routes, in the city of New York, viz.: To commence at the intersection of Tenth avenue and Fifty-ninth street; thence through and along Tenth avenue, with a double track, into West Twelfth street; thence through and along West Twelfth street, with a single track, to Greenwich street; thence, from West Twelfth street, through and along both West and Greenwich streets, southerly, with a single or double track upon each of said streets, to Battery place; thence through and along Battery place to State street, with double track; thence through and along State street, with single track, to Whitehall street; thence through and along Whitehall street, with double track, to South Ferry; returning through and along Whitehall street, with single track, from its intersection with State street, to Bowling Green; thence along the southerly side of Bowling Green, with single track, to connect with the double track in Battery place, with the right to construct, maintain, and use a double track, from West street, through and along Chambers street, to its intersection with Hudson street; also from the intersection of Tenth avenue and Fifty-ninth street, with double track, through and along Fifty-ninth street, to First avenue; thence through and along First avenue, with double track, to Twenty-third street; thence through and along Twenty-third street, with double track, to avenue A; thence through and along avenue A, with double track, to Fourteenth street; thence through and along Fourteenth street, with double track, to avenue D; thence through and along avenue D, with double track, to Houston street; thence through and along Houston street, with double track, to Mangin street; thence through and along Mangin street, with single track, to Grand street: thence through and along Grand street to Corlears street, with single track; thence through Corlears street to South street, with single track; thence through and along South street to Montgomery street, with single track ; thence through and along Montgomery street, with single track, to the junction of Front and South streets; thence through and along South street, with double track, to the junction of South and Front streets, at Roosevelt street; thence through and along South street to Old slip, with single track ; thence through and along Old slip to Water street, with single track ; thence through and along Water street to Whitehall street, with single track ; thence through and along Whitehall street to South street, with double track; thence through and along South street to Coenties slip, with single track ; thence through and along Coenties slip to Front street, with single track ; and also with single track from Old slip, through and along Front street to Whitehall street ; also a double track in Broad street, from Water street to South street ; also through and along Houston street from its intersection with avenue D by the track already named, to Goerck street; thence through and along Goerck street to Grand street, with single track ; thence through and along Grand street, with single track, to its intersection with Monroe street ; thence through and along Monroe street to Jackson street, with single track ; thence through and along Jackson street to Front street, with single track ; thence through and along Front street, with single track, to its intersection with South street at Montgomery street ; thence through and along South street, by the double track already named, to Front street, at the junction of South and Front streets, at Roosevelt street; thence through and along Front street to Old slip, and thence through and along Front street to Whitehall street, by the track already named; thence through and along Whitehall street, with single track, to South ferry, with the privilege of laying all necessary sidings, turnouts, connections, and switch for the proper working and accommodation of the said railroad, in any of the above-mentioned streets, and of connecting with, running on, or crossing all such other railroad tracks as may lie along or across any of said routes, streets, or avenues.

§ 2. Said railroad shall be constructed upon the most approved plan for the construction of city railroads, and the cars on the same shall run as often as the convenience of the public shall require, and shall be subject to such reasonable rules and regulations in respect thereto, in the transportation of passengers and freight in suitable cars, as the common council of the city of New York may, from time to time, by ordinance prescribe, and to the payment to the city of the same license fee annually, for each passenger car run thereon, as is now paid by other city railroads in said city; and no higher rate of fare shall be charged for the conveyance of passengers thereon than is now charged by the city railroads in said city, now chartered and constructed; and the said common council are hereby authorized and required to grant permission to the persons herein named, or their assigns, to construct, maintain, operate, and use said railroad in, upon, and along the several streets and avenues herein mentioned.

§ 3. In the construction, operation, and use of such railroad, should the said parties above named, or their assigns, deem it necessary or proper to run upon, intersect, or use any portion of any other railroad tracks now laid upon any of the streets or avenues above named, they are hereby authorized to run upon, intersect, and use the same; and in case they cannot agree with the owner or owners thereof respecting the compensation or payment to be made therefor, then the amount of such compensation or payment shall be ascertained and determined in the manner provided by subdivision six of the twenty-eighth section of the act entitled "An act to authorize the formation of rail road corporations and to regulate the same," passed April second, eighteen hundred and fifty ; and should any real estate or interest therein be required for the purpose of constructing said railroad on said route or routes, as above specified and authorized, for which the said persons above named, or their assigns, shall be unable to agree with the owner or owners for the use or purchase thereof, they may acquire the right to use, or title to the same, in the manner specified in the fourteenth, fifteenth, sixteenth, seventeenth, eighteenth, nineteenth, twentieth, and twenty-first sections of the said act of April second, eighteen hundred and fifty, except that, in any of the proceedings for any of the purposes authorized by this section, it shall not be necessary that the petition to the supreme court shall make any allegations of, or reference to any incorporation, capital, stock, surveys, or maps, or of the filing of any certificate of locations. But, in all cases, the use of said streets and avenues for the purposes of said railroad, as herein authorized, shall be considered one of the uses for which the mayor, aldermen, and commonalty of said city hold said streets and avenues.

§ 4. The mayor, common council, and the several officers of the corporation of the said city of New York, and the said corporation, are hereby prohibited from giving any assent to, or allowing any company claiming to derive authority, under the act entitled " An act to authorize the formation of railroad corporations and to regulate the same," passed April second, eighteen hundred and fifty, or act amendatory thereof, or in addition thereto, to construct any railroad in, or upon any or either of the said streets or avenues, and from doing any other act to hinder, delay, or obstruct the construction or operation of said rail road as herein authorized. And it is hereby made the duty of the said mayor, common council, and other officers, to do such acts, within their respective departments, as may be needful to promote the construction and protect the operation of the said railroad as provided in this law; any act or thing done in violation hereof shall be inoperative and void. All actions relating to, affecting, or arising under this act, or the authority herein given, shall be commenced in the supreme court of the first judicial district.

§ 5. All acts or parts of acts, inconsistent with the provision of this act, are hereby repealed, and declared to be inoperative so far as the same are applicable to this act.

§ 6. This act shall take effect immediately.

§ 7. The legislature may at any time modify, amend, or repeal this act.

CHAPTER 512.

AN ACT *to authorize the construction of a railroad in avenue D, East Broadway, and other streets and avenues of the City of New York.*

Passed April 17, 1860; notwithstanding the objections of the Governor.

The People of the State of New York, represented in Senate and Assembly, do enact as follows:

SECTION 1. John E. Devlin, William A. Hall, Cornelius Runkel, Bernard Smyth, Harry Clark, William A. Herring, William D. Marvin, John V. Coon, William P. Buckmaster, George L. Thomas, William N. Hays, James Murphy and their assigns, are hereby authorized to lay, construct, operate, and use a railroad with a double or single track, as hereinafter provided, and to convey passengers thereon for compensation, through, upon, and along the following streets and avenues, route or routes, in the city of New York, viz: Commencing on avenue D, at the northern extremity of the same ; thence through and along avenue D with a double track to Eighth street ; thence through and along Eighth street with a single track to Lewis street ; thence through and along Lewis street with a single track to Grand street ; thence through and along Grand street with a double track to East Broadway ; thence through and along East Broadway, Chatham square, Chatham street, and Park row, with a double track to Broadway ; also, from the corner of avenue D and Eighth street, through and along avenue D with a single track to Houston street ; thence through and along Houston street with a single track to Goerck street , thence through and along Goerck street with a single track to connect with a double track in Grand street, hereinafter provided for ; also connecting with the double track in East Broadway, through and along Canal street with a double track to the westerly side of Broadway : also connecting with the double track in Grand street at Lewis street, through and along Grand street with a double track to the Grand street ferry ; also, commencing at the northern extremity of avenue B, through and along avenue B, with a double track, to Clinton street; thence through and along Clinton street, with a double track, to connect with the track in East Broadway ; also connecting with the track in avenue B, through and along Tenth street and Eleventh street with single tracks to avenue D ; also, connecting with the double track in Canal street at Broadway, with a single track across Broadway to Lispenard street; thence, with a single track, through and along Lispenard street, to and across West Broadway to Beach street; thence through and along Beach street, with a single track, to Washington street; thence through and along Washington street, with a single track, to Battery place ; thence through and along Battery place, with a double track, to the Bowling Green at State street, also, connecting with the track in Battery place, through and along Greenwich street, with a double track, to the centre of Canal street; also, connecting with the track in Washington street at North Moore street, with a single track, through and along North Moore street and across West Broadway to Walker street, and thence through and along Walker street to, and to connect with, the double track in Canal street; also, connecting with the track in Washington street, through and along Washington street, with a single track, to the centre of Canal street, also, connecting with the double track in avenue D, through and along Fourteenth street, with a double track, to First avenue; thence through and along First avenue, with a double track, to Thirty-fourth street, thence through and along Thirty-fourth street, with a double track, to avenue A, and thence through and along avenue A, with a double track to, and to connect with, the double track in Fourteenth street, together with the necessary connections, turnouts, and switches for the proper working and accommodation of the road on the said route or routes.

§ 2. Said railroad shall be constructed on the most approved plan for the construction of city railroads, and shall be run as often as the convenience of passengers may require, and shall be subject to such reasonable rules and regulations in respect thereto as the common council of the city of New York may, from time to time, by ordinance prescribe; and to the payment to the city of the same license fee annually for each car run thereon, as is now paid by other city railroads in said city; and the said persons and their assigns are hereby authorized to charge the same rate of fare, for the conveyance of passengers on said railroad, as is now charged by other city railroads in said city.

§ 3. In the construction, operation, or use of such railroad, upon the route or routes above designated, should such persons above named, or their assigns, deem it necessary or proper to run upon, intersect, or use any portion of other railroad tracks now laid upon any of the streets or avenues above named, they are hereby authorized to run upon, intersect, and use the same, and in case they cannot agree with the owner or owners thereof respecting the compensation or payment to be made therefor, then the amount of such compensation or payment shall be ascertained and determined in the manner provided by subdivision six of the twenty-eighth section of the act entitled "An act to authorize the formation of railroad corporations, and to regulate the same," passed April second, eighteen hundred and fifty. And should any real estate or interest therein be required for the purpose of constructing said railroad on said route or routes, as above specified and authorized, for which the said persons above named, or their assigns, shall be unable to agree with the owner or owners for the use or purchase thereof, they may acquire the right to use, or title to the same, in the manner specified in the fourteenth, fifteenth, sixteenth, seventeenth, eighteenth, nineteenth, twentieth, and twenty-first sections of the said act of April second, eighteen hundred and fifty, except that, in any of the proceedings for any of the purposes authorized by this section, it shall not be necessary that the petition to the supreme court shall make any allegations of, or reference to any incorporation, capital stock, surveys, or maps, or of the filing of any certificate of location. But, in all cases, the use of said streets and avenues for the purposes of said railroad, as herein authorized, shall be considered a public use consistent with the uses for which the mayor, aldermen, and commonalty of said city hold said streets and avenues. The expense of constructing the tracks in Greenwich and Washington streets and Battery place, as herein provided, shall be borne equally by said persons or their assigns, and any company which is now or shall hereafter be authorized to construct tracks therein, and thereupon the said tracks shall be used in common by the said persons or their assigns and such company.

§ 4. The mayor, common council, and the several officers of the corporation of the said city of New York, and the said corporation are hereby prohibited from giving any assent to, or allowing any company claiming to derive authority under the act entitled "An act to authorize the

23

formation of railroad corporations, and to regulate the same," passed April second, eighteen hundred and fifty, or act amendatory thereof, or in addition thereto, to construct any railroad in or upon any or either of the said streets or avenues, and from doing any other act to hinder, delay, or obstruct the construction or operation of said railroad as herein authorized. And it is hereby made the duty of the said mayor, common council, and other officers, to do such acts, within their respective departments, as may be needful to promote the construction and protect the operation of said railroad, as provided in this law. Any act or thing done, in violation hereof, shall be inoperative and void. All actions relating to, or affecting or arising under this act, or the authority herein given, shall be commenced in the supreme court of the first judicial district. Nothing in this section contained shall be deemed or held to impair the rights of any railroad now in operation in said city.

§ 5. All provisions of law, inconsistent with this act, are hereby repealed.

§ 6. This act shall take effect immediately.

CHAPTER 513.

An Act to authorize the construction of a railroad in Seventh avenue, and in certain other streets and avenues of the city of New York.

Passed April 17th, 1860, notwithstanding the objections of the Governor.

The People of the State of New York, represented in Senate and Assembly, do enact as follows:

SECTION 1. John Kerr, Edward P. Cowles, Anthony J. Hill, Hugh Smith, John S. Hunt, Jacob Sharp, Thomas H. Tower, Peter B. Sweeney, John B. Babcock, Robert Marshall, John Kelly, Jacob Hays, and their assigns, are hereby authorized and empowered to lay, construct, operate, and use a railroad with a double or single track, as hereinafter provided, and to convey passengers thereon for compensation, through, upon, and along the following streets and avenues, route or routes, in the city of New York, viz.: Commencing on the Seventh avenue at the southern extremity of the Central Park; thence through and along the Seventh avenue with a double track to the old Bloomingdale road or Broadway, thence through and along the old Bloomingdale road or Broadway and Union place, with a double track to University place; thence through and along University place with a double track to Clinton place or Eighth street; thence through and along University place and Wooster street with a single track to Canal street; thence through and along Canal street with a single track to West Broadway, thence through and along West Broadway and College place with a single track to Barclay street; thence through and along Barclay street with a single track to Church street; thence through and along Barclay street with a double track to Broadway, also, connecting with the double track in Barclay street, through and along Church street with a single track to Canal street, thence through and along Canal street with a single track to Greene street, thence through and along Greene street with a single track to Clinton place or Eighth street.

thence through and along Clinton place or Eighth street with a single track to connect with a double track in University place; thence to the place of beginning; also connecting with the double track in Seventh avenue at Broadway, through and along Seventh avenue with a double track to Greenwich avenue; thence through and along Greenwich avenue with a double track to and across the Sixth avenue to Clinton place or Eighth street; thence through and along Clinton place or Eighth street with a double track to Macdougal street; thence through and along Macdougal street with a double track to Fourth street,'thence through and along Fourth street with a double track to Thompson street; thence through and along Thompson street with a double track to Canal street; thence through and along Canal street with a double track to West Broadway, thence through and along West Broadway with a double track to Chambers street; thence through and along West Broadway and College place with a single track to Barclay street; thence through and along Barclay street to Broadway; thence returning through Barclay street and Church street to Chambers street; thence through and along Chambers street to West Broadway, to connect with the track in said street, and by the aforesaid route to the place of beginning; also, connecting with the track in College place, through and along Park place with a double track to Broadway; also, connecting with the track in West Broadway, to and along Duane street with a single track to Church street, and thence through and along Duane street with a double track to Broadway; also, connecting with the track in Thompson street, through and along Broome street with a double track to Broadway, also, connecting with the track in

Union place at Fourteenth street, through and along Fourteenth street with a double track to Broadway, adjoining Union square; also connecting with the double track in Canal street at Thompson street, through and along Canal street with a double track to Varick street, and thence through and along Varick street with a double track to, and to connect with the track in West Broadway at Franklin street; together with the necessary connections, turnouts, and switches, for the proper working and accommodation of the said railroad, on the said route or routes.

§ 2. Said railroad shall be constructed on the most approved plan for the construction of city railroads, and shall be run as often as the convenience of passengers may require, and shall be subject to such reasonable rules and regulations in respect thereto as the Common Council of the city of New York may from time to time by ordinance prescribe; and to the payment to the city of the same license fee annually, for each car run thereon, as is now by other city railroads in said city; and the said persons and their assigns are hereby authorized to charge the same rate of fare, for the conveyance of passengers on said railroad, as is now charged by other city railroads in said city.

§ 3. In the construction, operation, or use of such railroad upon the route or routes above designated, should such persons above named, or their assigns, deem it necessary or proper to run upon, intersect, or use any portion of other railroad tracks now laid upon any of the streets or avenues above named, they are hereby authorized to run upon, intersect, and use the same, and in case they cannot agree with the owner or owners thereof respecting the com-

pensation or payment to be made therefor, then the amount
of such compensation or payment shall be ascertained and
determined in the manner provided by subdivision six of
the twenty-eighth section of the act entitled " An act to
authorize the formation of railroad corporations, and to regulate the same," passed April second, eighteen hundred and
fifty.   And should any real estate or interest therein be required for the purpose of constructing said railroad on said
route or routes, as above specified and authorized, for which
the said persons above named, or their assigns, shall o unable to agree with the owner or owners for the use   purchase thereof, they may acquire the right to use, or title to
the same in the manner specified in the fourteenth, fifteenth,
sixteenth, seventeenth, eighteenth, nineteenth, twentieth,
and twenty-first sections of the said act of April second,
eighteen hundred and fifty, except that, in any of the proceedings for any of the purposes authorized by this section,
it shall not be necessary that the petition to the supreme
court shall make any allegations of or reference to any
incorporation, capital stock, surveys or maps, or of the filing of any certificate of location.   But, in all cases, the use
of said streets and avenues for the purposes of said railroad,
as herein authorized, shall be considered a public use consistent with the uses for which the mayor, aldermen, and
commonalty of said city hold said streets and avenues.

§ 4. The mayor, common council, and the several officers
of the corporation of the said city of New York, and the
said corporation, are hereby prohibited from giving any
assent to, or allowing any company, claiming to derive authority under the act entitled " An act to authorize the
formation of railroad corporations, and to regulate the

same," passed April second, eighteen hundred and fifty, or
act amendatory thereof, or in addition thereto, to construct
any railroad in, or upon any or either of the said streets or
avenues, and from doing any other act to hinder, delay, or
obstruct the construction or operation of said railroad as
herein authorized. And it is hereby made the duty of the
said mayor, common council, and other officers, to do such
acts within their respective departments, as may be needful to promote the construction and protect the operation
of said railroad, as provided in this law. Any act or thing
done, in violation hereof, shall be inoperative and void. All
actions relating to, affecting, or arising under this act, or
the authority herein given, shall be commenced in the supreme court of the first judicial district. Nothing in this
section contained shall be deemed or held to impair the
rights of any railroad now in operation in said city.

§ 5. All provisions of law, inconsistent with this act, are
hereby repealed.

§ 6. This act shall take effect immediately.

CHAPTER 514.

AN ACT *to authorize the construction of a railroad in Fourteenth street, and in other streets and avenues of the City of
New York.*

Passed April 17, 1860; notwithstanding the objections of
the Governor.

The People of the State of New York, represented in
Senate and Assembly, do enact as follows ·

SECTION 1. Stephen R. Roe, John Stewart, Charles W. Lawrence, John Kennedy, James S. Hunt, Charles C. Clarke, John Fox, William Ravensteyn, William H. Peck, John C. Thompson, Thomas Ryan, Joseph S. Craig, and their assigns, are hereby authorized and empowered to lay, construct, operate, and use a railroad with a double or single track, as hereinafter provided, and to convey passengers thereon, for compensation, through, upon, and along the following streets and avenues, route or routes, in the city of New York, viz.: Commencing at the intersection of Fourteenth street with the Eleventh avenue; thence through and along Fourteenth street, with a double track, to Hudson street; thence through and along Hudson street, with a double track, to Troy street; thence through and along Troy street, with a single track, to Fourth street; thence through and along Fourth street, with a single track, to Macdougal street; thence through and along Macdougal street, with a single track, to Bleecker street; thence through and along Bleecker street, with a double track, to Crosby street; thence through and along Crosby street, with a double track, to Howard street; thence through and along Howard street, with a double track, to Elm street; thence through and along Elm street, with a double track, to Leonard street; thence through and along Elm street, with a single track, to Reade street; thence through and along Reade street, with a single track, to Centre street; thence through and along Centre street, Chatham street and Park row, with a double track, to Broadway; also, connecting with the double track, in Centre street, at Reade street, through and along Centre street, with a single track, to Leonard street; thence through and along Leonard street, with a single track, to connect with the double track, in

Elm street; also, connecting with the double track, in Hudson street, at Troy street, through and along Hudson street, with a single track, to the southerly end of Abingdon square and Bleecker street; thence through and along Bleecker street, with a single track, to Macdougal street, there to connect with the double track in Bleecker street; also, connecting with the double track in Park row, through and along Beekman street, with a single track, to South street; thence through and along South street, with a single track, to Fulton street; thence through and along Fulton street, with a single track, to William street; thence through and along William street, with a single track, to Ann street, thence through and along Ann street, with a single track, to connect with the double track in Park row at Broadway; also, connecting with the double track in Elm street, through and along Canal street, with a double track, to Broadway; also, with a double track connecting with the double track in Fourteenth street, through and along the Eleventh and Twelfth avenues to Thirty-second street; also, connecting with the double track in Canal street at Elm street, through and along Canal street with a double track to the Bowery, thence through and along the Bowery and New Bowery, with a double track, to Pearl street; thence through and along Pearl street, with a double track, to Peck slip; thence through and along Peck slip, with a double track, to South street· thence through and along South street, with a double track, to the Fulton ferry; thence through and along Fulton street, with a double track, to Water street thence through and along Water street, with a double track, to connect with the said double track in Peck slip, together with the necessary connections, turnouts, and switches for the proper working and accommodation of the road on the said route or routes.

§ 2. Said railroad shall be constructed upon the most
approved plan for the construction of city railroads, and
shall be run as often as the covenience of passengers may
require, and shall be subject to such reasonable rules and
regulations in respect thereto as the common council of the
city of New York may, from time to time, by ordinance
prescribe; and to the payment to the city of the same
license fee annually, for each car run thereon, as is now paid
by other city railroads in said city; and the said persons and
their assigns are hereby authorized to charge the same rate
of fare, for the conveyance of passengers on said railroad, as
is now charged by other city railroads in said city.

§ 3. In the construction, operation, or use of such railroad, upon the route or routes above designated, should
such persons above named, or their assigns, deem it necessary or proper to run upon, intersect, or use any portion of
other railroad tracks now laid upon any of the streets or
avenues above named, they are hereby authorized to run
upon, intersect, and use the same, and in case they cannot
agree with the owner or owners thereof respecting the compensation or payment to be made therefor, then the amount
of such compensation or payment shall be ascertained and
determined in the manner provided by subdivision six of
the twenty-eighth section of the act entitled " An act to
authorize the formation of railroad corpora'ions and to regulate the same," passed April second, eighteen hundred and
fifty. And should any real estate or interest therein be required for the purpose of constructing said railroad on said
route or routes, as above specified and authorized, for which
the said persons above named, or their assigns, shall be
unable to agree with the owner or owners for the use or

purchase thereof, they may acquire the right to use, or
title to the same, in the manner specified in the fourteenth,
fifteenth, sixteenth, seventeenth, eighteenth, nineteenth,
twentieth, and twenty-first sections of the said act of April
second, eighteen hundred and fifty, except that, in any of
the proceedings for any of the purposes authorized by this
section, it shall not be necessary that the petition to the
supreme court shall make any allegations of, or reference to
any incorporation, capital stock, surveys, or maps, or of
the filing of any certificate of location. But, in all cases
the use of said streets and avenues for the purposes of said
railroad, as herein authorized, shall be considered a public
use consistent with the uses for which the mayor, aldermen,
and commonalty of said city hold said streets and avenues.

§ 4. The mayor, common council, and the several officers
of the corporation of the said city of New York, and the
said corporation, are hereby prohibited from giving any
assent to, or allowing any company, claiming to derive
authority under the act entitled "An act to authorize the
formation of railroad corporations, and to regulate the
same," passed April second, eighteen hundred and fifty, or
act amendatory thereof, or in addition thereto, to construct
any railroad in, or upon any or either of the said streets or
avenues, and from doing any other act to hinder, delay, or
obstruct the construction or operation of said railroad as
herein authorized. And it is hereby made the duty of the
said mayor, common council, and other officers to do such
acts, within their respective departments, as may be needful to promote the construction and protect the operation
of said railroad, as provided in this law. Any act or thing
done, in violation hereof, shall be inoperative and void. All

actions relating to, affecting, or arising under this act, or the authority herein given, shall be commenced in the supreme court of the first judicial district. Nothing in this section contained shall be deemed or held to impair the rights of any railroad now in operation in said city.

§ 5. All provisions of law, inconsistent with this act, are hereby repealed.

§ 6. This act shall take effect immediately.

CHAPTER 515.

AN ACT to authorize the construction of a railroad in Tenth avenue, Forty-second street, and certain other avenues and streets of the city of New York.

Passed April 17, 1860; notwithstanding the objection of the Governor.

The People of the State of New York, represented in Senate and Assembly, do enact as follows :

SECTION 1. John T. Conover, Moses Ely, Matthew T. Brennan, Truman Smith, Rufus F. Andrews, Bloomfield Usher, Justin D. White, John M. Miller, Elijah B. Holmes, Leonard W. Brainard, Junior, Delos De Wolf, Thomas Black, and their assigns, are hereby authorized and empowered to lay, construct, operate, and use a railroad with a double or single track, as hereinafter provided, and to convey passengers thereon for compensation, through, upon, and along the following streets and avenues in the city of New York, viz.: Commencing at the ferry, at the western extremity of Forty-second street; thence through and along Forty-second street, with a double track, to Tenth avenue; thence through and along Tenth avenue, with a double track, to Thirty-fourth street, thence through and along Thirty-fourth street, with a double track, to Broadway, thence through and along Broadway, with a double track, to Twenty-third street; thence through and along Twenty-third street, with a double track, to Fourth avenue; thence through and along Fourth avenue and Union place, with a double track, to Fourteenth street; thence through and along Fourteenth street, with a double track, to avenue A; thence through and along avenue A, with a double track, to Second street; thence through and along avenue A, with a single track, to First street; thence through and along First street and Houston street, with a single track, to Cannon street; thence through and along Cannon street, with a single track, to Grand street; thence through and along Grand street, with a single track, to Goerck street; thence through and along Grand street, with a double track, to Grand street ferry, at the foot of Grand street, East river; thence returning through and along Grand street to Goerck street; thence through and along Goerck street, with a single track, to Houston street; thence through and along Houston street and Second street, with a single track, to, and to connect with, the double track in avenue A, and thence along the aforesaid route to the place of beginning; together with the necessary connections, turnouts and switches, for the proper working and accommodation of the road on the said route.

§ 2. Said railroad shall be constructed on the most approved plan for the construction of city railroads, and shall be run as often as the convenience of passengers may require, and shall be subject to such reasonable rules and regulations in respect thereto as the common council of the city of New York may from time to time by ordinance prescribe; and to the payment to the city of the same license fee annually, for each car run thereon, as is now paid by other city railroads in said city; and the said persons and their assigns are hereby authorized to charge the same rate of fare for the conveyance of passengers on said railroad as is now charged by other city railroads in said city.

§ 3. In the construction, operation, or use of such railroad upon the route or routes above designated, should such persons above named, or their assigns, deem it necessary or proper to run upon, intersect, or use any portion of other railroad tracks now laid upon any of the streets or avenues above named, they are hereby authorized to run upon, intersect, and use the same, and in case they cannot agree with the owner or owners thereof respecting the compensation or payment to be made therefor, then the amount of such compensation or payment shall be ascertained and determined in the manner provided by subdivision six of the twenty-eighth section of the act entitled "An act to authorize the formation of railroad corporations, and to regulate the same," passed April second, eighteen hundred and fifty. And should any real estate or interest therein be required for the purpose of constructing the said railroad on the said route or routes, as above specified and authorized, for which the said persons above named, or their assigns, shall be unable to agree with the owner or owners for the use or purchase thereof, they may acquire the right to use, or title to the same, in the manner specified in the fourteenth, fifteenth, sixteenth, seventeenth, eighteenth, nineteenth, twentieth, and twenty-first sections of the said act of April second, eighteen hundred and fifty, except that, in any of the proceedings for any of the purposes authorized by this section, it shall not be necessary that the petition to the supreme court shall make any allegations of, or reference to any incorporation, capital stock, surveys or maps, or of the filing of any certificate of location. But, in all cases, the use of said streets and avenues for the purposes of the said railroad, as herein authorized, shall be considered a public use consistent with the uses for which the mayor, aldermen, and commonalty of said city hold said streets and avenues.

§ 4. The mayor, common council, and the several officers of the corporation of the said city of New York, and the said corporation, are hereby prohibited from giving any assent to, or allowing any company, claiming to derive authority under the act entitled " An act to authorize the formation of railroad corporations, and to regulate the same," passed April second, eighteen hundred and fifty, or act amendatory thereof, or in addition thereto, to construct any railroad in, or upon any or either of the said streets or avenues, and from doing any other act to hinder, delay, or obstruct the construction or operation of said railroad, as herein authorized. And it is hereby made the duty of the said mayor, common council, and other officers, to do such acts, within their respective departments, as may be needful to promote the construction and protect the operation of said railroad, as provided in this law. Any act or thing done, in violation hereof, shall be inoperative and void. All actions relating to, affecting, or arising under this act, or the authority herein given, shall be commenced in the supreme court of the first judicial district. Nothing in this section contained shall be deemed or held to impair the rights of any railroad now in operation in said city.

§ 6. All provisions of law, inconsistent with this act, are hereby repealed.

§ 7. This act shall take effect immediately.

www.ingramcontent.com/pod-product-compliance
Lightning Source LLC
Chambersburg PA
CBHW021807190326
41518CB00007B/481